Climate and Energy Decoded

A Realistic Overview of Climate Change, Renewable Energy & Low Carbon Transition

Tushar Choudhary, Ph.D.

HopeSpring Press

Published by HopeSpring Press
Rosenberg, Texas

Paperback ISBN: 979-8-9864358-0-0
eBook ISBN: 979-8-9864358-1-7

Library of Congress Control Number: 2022915147

Cover Design Credit: Rajeshree Choudhary

To my amazing wife and best friend,
Rajeshree Choudhary.

Table of Contents

PREFACE 11

INTRODUCTION 13

A Realistic Perspective is Needed 13

Book Purpose 13

Key Elements 14

1. CLIMATE CHANGE SCIENCE 15

Weather vs. Climate 15

Climate Monitoring 15
- Tree rings 16
- Ice cores 16

Climate Change 17

Natural Causes of Climate Change 17
- Solar variation 18
- Orbital changes 18
- Volcanic eruptions 18

Greenhouse gases 19
- Carbon-dioxide (CO_2) 20
- Methane 20
- Nitrous oxide 21
- Fluorinated gases 21

Human-Caused Climate Change 21
- Background 22
- Evidence 22
- Effects 24

2. HISTORY OF CLIMATE CHANGE 27

History of Climate Change Science 27
- Warming effect of greenhouse gases 27
- Climate monitoring 28
- Climate of the distant past 30

Understanding the severity of human-caused climate change 30

International Efforts **31**
Intergovernmental panel on climate change (IPCC) 32
Kyoto Protocol 32
Paris Agreement 33

3. CLIMATE CHANGE DEBATE **35**

Is Human-Caused Climate Change a Serious Problem? **35**
What is the conclusion from the scientific efforts? 35
Is the scientific community colluding? 36

Why the Significant Misunderstanding? **36**
Inappropriate messengers: Polarizing politicians 37
Inappropriate messengers: Celebrities 37
Ineffective messaging 38

4. A BRIEF HISTORY OF ENERGY **41**

Energy Use before Fossil Fuels **41**

Fossil Fuels Era **41**
Transition to coal 42
Transition to liquid petroleum fuels (Crude Oil) 43
Transition to natural gas 44

Key Lessons from Prior Energy Transitions **45**

5. INTRODUCTION TO LOW-CARBON ENERGY **47**

Key Characteristics **47**
Key Characteristic 1: Cost 47
Key Characteristic 2: Convenience 48

Approach for Discussing the Low-Carbon Energy Options **48**

6. LOW-CARBON POWER OPTIONS **51**

Solar Power **52**
Technology Basics 52
Utility Solar: Advantages and Challenges 53
Residential Solar: Advantages and Challenges 55

Solar Power with Energy Storage **55**

Technology Basics 56
Advantages and Challenges 56

Wind Power 57
Technology Basics 58
Onshore Wind Power: Advantages and Challenges 58
Offshore Wind Power: Advantages and Challenges 59

Wind Power with Energy Storage 60

Hydropower 60
Technology Basics 60
Advantages and Challenges 61

Nuclear Power 61
Technology Basics 62
Advantages and Challenges 62

Biomass Power 63
Technology Basics 63
Advantages and Challenges 63

Fossil Fuel Plants with Carbon Capture & Storage (CCS) 64
Technology Basics 64
Advantages and Challenges: Natural gas power plants with CCS 65
Advantages and Challenges: Coal power plants with CCS 65

Geothermal Power 65
Technology Basics 66
Advantages and Challenges 67

7. LOW-CARBON TRANSPORTATION OPTIONS 69

Battery Electric Vehicles (BEVs) 69
Technology Basics 70
Advantages and Challenges: Light duty vehicles 71
Advantages and Challenges: Road Freight, Aviation and Shipping 72

Fuel Cell Electric Vehicles 73
Technology Basics 73
Advantages and Challenges: Light duty vehicles category 74
Advantages and Challenges: Road Freight, Aviation and Shipping 74

Advanced Biofuels 74
Technology Basics 75
Advantages and Challenges: Light duty vehicles 75
Advantages and Challenges: Road Freight, Aviation & Shipping 76

8. THE DEBATE AROUND ENERGY OPTIONS 77

Common Issues for Low-Carbon Energy Options 77
What are the common flaws related to the discussions about low-carbon technologies? 77

The Debate Around Renewable Power 79
Are renewable technologies such as solar and wind not viable? 79
Can the cost of solar and wind be directly compared with fossil fuel and other dispatchable technologies? 80
Why the delay, when some countries already generate most of their electricity from renewable sources? 81
Can cost of solar and wind power become lower than fossil fuel power for 24X7 electricity production? 83
Don't the learning curves indicate that solar and wind costs will soon become lower than fossil fuel power? 87
Why is there so much debate about green hydrogen? 88
How much subsidies have renewables received? 89

The Debate Around Fossil Fuel Technologies 90
Is it appropriate to lump natural gas power plants with coal power plants? ... 90
Do fugitive emissions cancel the advantages of natural gas power? 91
What does science say about the impact of fracking on water quality? 92
What impact do fossil fuels have on air pollution? 93
Why have fossil fuel energy costs fluctuated widely over the past decades?. 94
Do fossil fuels receive several trillion dollars in subsidies each year? 94

The Debate Around Transportation Solutions 95
Apart from electric vehicles, what options can reduce greenhouse gases from light duty vehicles? 95
Are electric vehicles currently a very effective solution for addressing climate change? 97
Are we close to resolving the convenience challenge of BEVs? 100

The Debate Around Direct CO_2 Capture Technology 103
Can CO_2 capture from air become a wide-scale solution in the foreseeable future? 103

9. KEY ISSUES AROUND CLIMATE AND ENERGY 105

Human-Caused Climate Change: Key Issues 105
Should we ignore the conclusions from climate research because certain past claims have been unreliable? 105
Should we ignore climate change since deaths from climate disasters have decreased drastically? 106
How can climate change be a serious problem when it is yet to be understood perfectly? 107

What is the consensus from climate experts about the future impacts of climate change? 108
What is the estimated impact of climate change on the economy? 109
Is climate change mitigation the most important issue for humans currently? 111
How long did it take to definitively understand that CO_2 from fossil fuels was a serious problem? 112
Why did the scientific community take so long to definitively understand the fossil CO_2 impact? 115
Will solar and wind not have any severe environmental impact when deployed at a wide-scale? 116
Is it appropriate for media to make sensational claims based on isolated peer reviewed articles? 119

Energy Transition: Key Issues 121
What sectors are the major contributors to global greenhouse gas emissions? 121
What countries are the major contributors to greenhouse gas emissions? 121
What are the key challenges related to electrification? 123
What is the cost for a global low carbon transition? 125
What are key concerns about the critical minerals required for a low-carbon energy transition? 126
Is recycling an easy solution for the large materials intensity of low-carbon technologies? 127
How does the amount of energy consumed by a country impact its economic well-being? 129
What are the key challenges associated with a low carbon energy transition? 130
Why is there so much unrealistic optimism about the low-carbon transition? 132
Is there consensus amongst global energy experts about the path forward for the energy transition? 134

10. THE BIG PICTURE DISCUSSION 137

Summary of Crucial Aspects 137
Climate change impact 137
Acceptance of human-caused climate change being a serious problem 138
Options for the power sector 138
Options for the transportation sector 141
Learnings from fossil fuel technologies and implications 143
Low-carbon energy transition 143

Requirements for efficiently addressing climate change 145

Path Forward Framework 146
Generic framework 146

Country-specific framework 152

Concluding Remarks 154

GLOSSARY & UNITS 157

APPENDIX 1: LIST OF ORGANIZATIONS WHICH SUPPORT THAT CLIMATE CHANGE IS A SERIOUS PROBLEM. 159

ABOUT THE AUTHOR 163

REFERENCES AND NOTES 165

Preface

Over the years, I have discussed climate and energy with people from diverse backgrounds. I have been both fascinated and troubled by these discussions. Fascinated because of the extreme contradiction in opinions and troubled because most opinions were based on poor information. I have seen a similar troubling trend on social media platforms.

I have traced the problem to the publications available on this topic. Broadly, three types of publications are available:

- Publications belonging to the first type have too much information. The overload makes it difficult to tease out crucial information.
- Publications belonging to the second type provide localized information and exclude crucial aspects. Such publications are misleading.
- Publications belonging to the third type are opinion-based and not guided by science.

I have written this book to communicate crucial information about climate change and energy in a concise and balanced manner.

People on the far extremes might dislike a balanced discussion. Some could think that the book is against fossil fuels, while others could think that the book is pessimistic about renewable technologies.

I am hopeful that most will recognize the book for what it is–a balanced discussion that provides a realistic understanding about climate change and the energy transition.

Many authors and commentators who write about this topic have strong financial ties to fossil fuels or green energy. Financial ties include employment or funding for academic research or funding for special interest promotion. I am not financially dependent on fossil fuels or green energy.

Tushar V. Choudhary, Ph.D.
Houston, Texas

Introduction

"The time to repair the roof is when the sun is shining."–***John F Kennedy***

A Realistic Perspective is Needed

The statements below represent two distinct sentiments around climate change and energy:

- Human actions are drastically changing the climate. The path forward should be to immediately shift from fossil fuels to renewable energy.
- Human actions have a minor impact on the climate. The path forward should be to maintain the status quo.

Clearly, there is a substantial disagreement.

Worldwide, local, and national governments are in the early stages of developing energy policies to address climate change. Over time, these polices will have a large impact on every person on the planet.

For a good outcome, the global population will need to support robust policies and reject poor policies.

How to distinguish between a robust and poor policy? This requires a realistic perspective about climate change and energy.

Book Purpose

A realistic perspective requires access to crucial information.

What is crucial information? It is everything one needs to know to make informed decisions.

For example, crucial information for selecting a surgeon for a life-critical surgery includes the following: How many similar surgeries has the surgeon undertaken? What is the surgeon's success rate? Is the surgeon accessible from a location and cost viewpoint?

This information includes everything one needs to know to select an appropriate surgeon. The surgeon's personal life and political affiliation is not important. In fact, such information can lead to poor decisions.

A balanced discussion is an unbiased presentation of all the crucial information. If people have extreme opinions, a balanced discussion will cause discomfort to some people on both sides. But a balanced discussion cannot be avoided. It is essential for developing a practical understanding about complex issues.

For example, a balanced discussion about politics in the United States will involve an unbiased assessment of both, the Republican and Democratic parties, i.e., it will consider strengths and weaknesses of both parties.

The purpose of this book is to provide crucial information about climate change and energy via a concise and balanced discussion.

Key Elements

- The book includes discussions about climate change, the history of energy, low carbon energy solutions, the debates around key issues and the big picture.
- Climate discussions cover climate change science, the history of climate change, and the debate around human-caused climate change.
- History of energy discussions cover how and why the previous energy transitions occurred.
- Low carbon energy discussions address the key issues and commonly encountered myths.
- The big picture discussion provides a framework for efficiently addressing climate change.
- The book is focused on practical information and credible data–such as data from governmental organizations.
- The discussions are simplified for concise and easy to understand messaging.

1. Climate Change Science

"Scientists work to fill the gaps in human knowledge and to build a theory that can explain observations of the world."–**Nature Journal**

A productive debate about the path forward requires a basic knowledge of climate change science. Crucial aspects are discussed herein.

Weather vs. Climate

Weather and climate are sometimes used interchangeably. Since this causes confusion, weather must be clearly distinguished from climate.

Weather is a short-term condition–something that's happening to the atmosphere over a short period at a given location. Factors such as air pressure, temperature, humidity, and wind speed can impact the atmosphere.

Local weather can change frequently from minute-to-minute, hour-to-hour, and day-to-day. For example, New York city can be sunny with a high temperature of 30° C one day, but rainy with a high temperature of 15° C the next day.

Weather has limited predictability and is virtually unpredictable beyond a couple of weeks[1].

Climate is how the atmosphere behaves over a long term at a certain location. It can be defined as the average of the weather over a certain period for a given location.

The quote by Robert Heinlein summarizes the difference between climate and weather. *"Climate is what we expect, weather is what we get."*

Climate Monitoring

Frequent and accurate measurements are necessary to understand the climate. Scientists have developed an array of monitoring tools to enable accurate and frequent climate measurements.

Meteorological stations, satellites and ocean buoys are used to monitor the weather and climate.

- Meteorological stations monitor temperature, rainfall, snow-depth and more.
- Satellites monitor clouds, storms, snow cover, volcanic activity, atmospheric ozone, and sea ice.

- Buoys monitor surface water and deep ocean temperatures.

The scientific community has access to data from a vast network of monitoring tools. For example, the monitoring network of the World Meteorological Organization includes over ten thousand surface weather stations, thousand upper-air stations, seven thousand ships, over thousand buoys, hundreds of weather radars, three thousand specially equipped commercial aircraft, thirty meteorological and two hundred research satellites[2].

These monitoring tools have been providing accurate climate measurements from the past several decades.

How about climate measurements from the distant past? For this, scientists use the data that has been preserved in tree rings, ice cores, corals, and ocean sediments from hundreds or even millions of years[3]. The study of past climate–aka paleoclimatology–has played a major role in improving the understanding of climate science[4].

Tree rings and ice cores are discussed below as examples of how scientists study the past climates.

Tree rings

Trees can live up to thousands of years and are sensitive to local climate conditions. The climate information is stored in concentric rings which can be observed from a top view of the tree stump. The tree rings grow wider and darker in the years that have hotter temperatures and higher rainfall. On the other hand, the tree rings are thinner in the years that have cooler temperatures and lower rainfall. Thus, trees are a storehouse of climate information.

Tree rings have provided information about annual changes in precipitation and temperature that have occurred over thousands of years[5].

Ice cores

Ice cores can be imagined as a collection of time capsules, wherein information about different time periods has been stored from several hundred thousand years[6].

Specifically, ice cores are samples that are drilled from glaciers to obtain climate data from ancient times. Glaciers form via the layered accumulation of snow. Each layer of snow has different texture and chemistry–as determined by the conditions prevailing at the time of the snow fall. The weight from the top layers of snow compresses the bottom layers and converts it to ice over time.

The ice contains particulates, dissolved chemicals and air bubbles captured by the falling snow. Consequently, each layer of ice in the glacier contains information that is specific to the period corresponding to the snow fall. The ice accumulates over seasons and years and contains information such as temperature and chemical composition of the atmosphere[7].

The different sections of the ice core correspond to different seasons and years. For example, the youngest ice is located at the top. Scientists analyze the ice at different locations along the ice core. They use the information to recreate past climate records. Such studies provide climate information over several hundred thousand years. This information has been very valuable for understanding climate change.

Climate Change

Climate change refers to long term changes in average weather patterns. It can occur on a local, regional or a global scale. For example, scientists can monitor the dryness of summer by studying the rainfall, water body levels, and satellite data for a given area. If the data over multiple summers indicates that the summers have become significantly drier than normal, it is an indication of climate change for the area. Examples of climate change on a global scale include the beginning and end of the ice ages.

Climate can change because of natural or human causes[8].

Natural Causes of Climate Change

To understand human-caused climate change, it is essential to first understand the natural causes.

Fortunately, decades of systematic research have provided a robust understanding about the natural causes of climate change. Natural causes have been changing earth's climate since the beginning of time. These natural causes include earth's orbital changes, solar variations, volcanic eruptions, ocean currents, meteorite impacts, internal variability such as El Nino, and plate tectonics[9,10]. Three major natural causes are discussed below.

Solar variation

The sun is earth's primary source of energy. Expectedly, it can influence the climate.

For example, a solar cycle can impact the solar output. The sun's magnetic field goes through a cycle approximately every eleven years. The cycle affects the activity on the surface of the sun which can cause minor variations in solar radiation.

The recent notable event related to solar variation occurred about four centuries ago and lasted for several decades[11]. The event was triggered by a decrease in solar radiation and cooling from volcanic activity[12].

How is the impact from solar variation estimated? By measuring the solar output.

Satellites have been used to monitor the solar output for the last forty years. The monitoring has revealed a very low variation in solar radiation[13]. This observation is indicative of a minimal contribution from solar variation to climate change over the past few decades.

Orbital changes

Orbital changes involve the changes in the position of earth with respect to the sun. Specifically, they include the change in the shape of earth's orbit, the tilt of earth's axis and the wobbling of earth's axis[14].

The orbital changes influence the climate by changing the amount of solar radiation that reaches earth. The resulting climate fluctuations occur over ten-thousands to hundreds-of-thousands of years.

The orbital changes are considered as one of the triggers for the beginning and the ending of the ice ages. The last ice age ended approximately twenty thousand years ago[15].

The climate impacts from the orbital changes are gradual and occur over a very long period. Thus, they are only noticeable over thousands-of-years.

Volcanic eruptions

Eruption of a volcano results in the release of volcanic ash, sulfur compounds, greenhouse gases and other debris in the atmosphere. Certain components such as volcanic ash and sulfur compounds block some of the sunlight and cool the climate. The effect is temporary–for a year or so–until the components remain in the atmosphere.

The greenhouse gases emissions from the eruptions cause a warming effect. This effect is small because of the low emission levels.

Overall, volcanic eruptions cause a short-term cooling effect. For example, the 1991 Mount Pinatubo volcanic eruption caused a 0.5° C drop in the global temperature for about fifteen months[16].

Greenhouse gases

Greenhouse gases play a crucial role in defining earth's temperature.

The importance of greenhouse gases can be understood from basic physics. Solar radiation reaches earth. Some of that energy is reflected into space, while some is absorbed and released as heat energy. The greenhouse gases in the atmosphere block some of this heat energy from escaping by absorbing and reflecting it in all directions. This partial blanketing effect from the greenhouse gases is responsible for the comfortable temperature on earth. Without this effect, the earth's average surface temperature would have been a hostile -18° C[17].

Water vapor, a condensable gas, is the major greenhouse component in the atmosphere. The minor components include carbon dioxide, methane, nitrous oxide, and fluorinated gases.

Notably, the minor greenhouse components control earth's temperature. This is because carbon dioxide, methane, nitrous oxide, and fluorinated gases are non-condensable, long-lived gases. Further emission of these gases increases their long-term atmospheric content. This increases the blanketing effect, which in turn increases earth's temperature.

This principle is also evidenced from the very high surface temperatures observed for some planets. Venus, whose atmosphere consists mainly of CO_2, has a mean surface temperature of 464° C[18].

An increase in the long-lived greenhouse gases increases the earth's temperature. This in turn increases the water vapor content. The greenhouse effect from the additional water vapor content further increases earth's temperature. Such an effect is described as positive feedback, i.e., a feedback response that increases the temperature.

If there is no increase in earth's temperature from the long-lived greenhouse gases, there will no increase in the water vapor content. Thus, contrary to the popular myth, water vapor does not control the rise in earth's temperature[19]. Consequently, the focus is on the long-lived greenhouse gases.

The key characteristics of the long-lived gases are a) heat trapping ability and b) atmospheric lifetime[20]. These characteristics can be stated in terms of a 100-year global warming potential.

The 100-year global warming potential is the heat energy that can be absorbed by a greenhouse gas relative to the heat energy that can be absorbed by the same amount of carbon dioxide over a period of hundred years.

Carbon-dioxide (CO_2)

CO_2 is released via human activities such as burning fossil fuels, deforestation, and cement production. Natural process such as volcanic eruptions and respiration also release CO_2. Energy production is the largest contributor to CO_2 emissions from human activity.

Oceans and plants absorb approximately half of the human-emitted CO_2[21]. A large fraction of the other half remains in the atmosphere for hundreds of years.

The 100-year global warming potential for CO_2 is 1.0 because it is the reference greenhouse gas[22].

CO_2 emissions are overwhelmingly larger than the other long-lived greenhouse gases. The total impact of a greenhouse gas depends on its global warming potential as well as how much is emitted. To account for both factors, the greenhouse gases are expressed in terms of CO_2 equivalent tons (1 ton = 1000 Kg)[23].

The CO_2 equivalent tons for each greenhouse gas are estimated by multiplying the amounts of emission of that greenhouse gas with its global warming potential. Thus, the use of CO_2 equivalent tons allows comparison of the relative contribution from the different greenhouse gases.

CO_2 emissions represent about 73% of the total human-caused emissions of greenhouse gases when considered in terms of CO_2 equivalent tons, i.e., based on its global warming potential and emission amounts[24].

Methane

Methane is released from human activities and natural sources. Human activities contribute to 60% of the total methane emissions[25,26]. Majority of the methane emissions from human activities are from the agriculture sector. Less than 25% of the total methane emissions are from the energy sector.

Methane has an atmospheric lifetime of twelve years. The 100-year global warming potential of methane is 28. This means that methane is 28 times more effective than CO_2 for trapping heat on an equivalent mass basis over a 100-year period[27]. Despite its higher effectiveness for trapping heat, the overall contribution to warming from methane is substantially lower than CO_2. This is because methane emissions are much lower than CO_2 emissions.

Methane emissions represent about 19% of the total human-caused greenhouse gas emissions[28].

Nitrous oxide

Nitrous oxide is released during agriculture and industrial activities as well as from the burning of fossil fuels and other waste. Agriculture is the largest contributor to nitrous oxide emissions from human activity.

Nitrous oxide has a lifetime of over hundred years and a 100-year global warming potential of 265. Nitrous oxide emissions are much lower than methane.

Nitrous oxide emissions represent about 5% of the total human-caused greenhouse gas emissions[29].

Fluorinated gases

Fluorinated gases include components such as chloroflurohydrocarbons, hydrofluorocarbons, perfluorocarbons, hydrochlorofluorocarbons, and sulfur hexafluoride. These gases are emitted from industrial activities and household applications such as refrigerant use.

Their lifetimes range from a few weeks to several thousand years. Most of the gases have very high 100-year global warming potentials, in the range of thousands to tens-of-thousands. The emissions of fluorinated gases are much lower than other long-lived greenhouse gases.

Fluorinated gases represent about 3% of the total human-caused greenhouse gas emissions[30].

Human-Caused Climate Change

Many politicians and celebrities have strong opinions about human-caused climate change. They frequently express these opinions and disguise them as facts. The problematic messaging around this topic is discussed in a subsequent chapter.

Herein, the focus is on the collective conclusions drawn from thousands of peer-reviewed scientific studies about human-caused climate change[31,32].

Background

Human-caused climate change is caused by activities such as burning fossil fuels, clearing forests, cement production and agriculture. These human activities add large amounts of greenhouse gases to those naturally existing in the atmosphere. This raises the greenhouse gas levels in the atmosphere beyond optimum levels and affects the climate. Currently, human activities produce 52 billion tons of greenhouse gases each year.

The burning of fossil fuels is by far the largest contributor. This is expected because a) the burning of fossil fuels produces CO_2 and b) the global population consumes enormous amounts of fossil fuels for energy production.

The cumulative CO_2 emissions from fossil fuels have increased from 0.01 billion metric tons in 1751 to over 2000 billion tons currently[33,34]. For reference, 1 billion ton = 1000,000,000,000 kg. Studies have revealed that about half of the emitted CO_2 remains in the atmosphere[35].

In the beginning, the scientific discussion was mainly focused on global warming. In recent decades, the discussion has extended to climate change, which includes several other impacts.

Evidence

Climate monitoring tools have provided high quality information from the recent past and ancient times. This climate information is robust evidence for human-caused climate change.

Over the past decades, human activities have led to enormous emissions of greenhouse gases. Correspondingly, the global monitoring stations have recorded a large increase in the atmospheric content of the different greenhouse gases. For example, the atmospheric CO_2 content has increased by 12%, methane by 7% and nitrous oxide by 5% over just the past couple of decades[36].

The atmospheric CO_2 content has been monitored rigorously over the decades. The content has increased from 317 ppm to 412 ppm from 1960 to 2020[37]. Furthermore, ice core studies have shown that the current atmospheric CO_2 levels are markedly higher than they have been in the past eight hundred thousand years[38]. Specifically, the atmospheric CO_2

levels had remained between 160 and 300 ppm for the last eight hundred thousand years before the rapid increase from the burning of fossil fuels.

The greenhouse gases have significantly increased the average surface temperature of earth. The evidence can be presented in terms of a global average surface temperature anomaly for each year (**Figure 1.1**)[39].

The data shows how much warmer or colder any particular year has been compared to long-term average temperature. Specifically, the Figure provides the temperature difference (anomaly) for each year relative to the average temperature from 1951 to 1980. For the year 2020, a temperature anomaly of about 1°C is observed relative to 1951-1980 average temperature. Notably, nineteen of the twenty hottest years over the last one hundred and forty years have occurred since the beginning of this century.

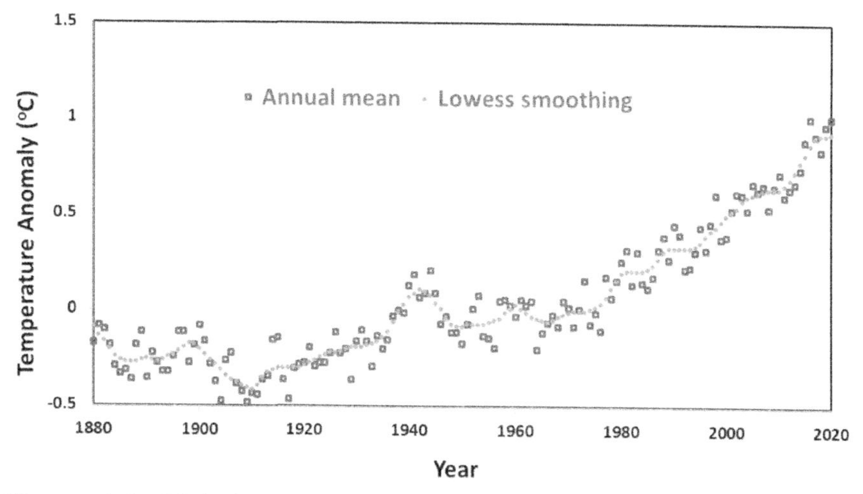

Figure 1.1: Global average surface temperature time-series. Change in global surface temperature relative to the average temperature from 1951 to 1981. Source: NASA[40].

Extensive studies have confirmed that natural causes cannot explain the observed temperature anomalies. Paleoclimatology studies have revealed that natural climate cycles occur at extremely slow rates[41]. Consequently, the human-caused change can be distinguished from natural causes such as orbital changes because of its rapid speed.

Also, satellite measurements of the solar output have showed very little variation over the past several decades[42]. This has demonstrated that solar variation has not contributed to the temperature rise.

Models are powerful tools for studying complex systems. Hence, researchers have paid special attention to developing climate models. They have used the climate models to explain the temperature anomalies discussed earlier[43].

Each modeling study found that the temperature anomalies could only be explained by allowing for a strong influence of greenhouse gases on temperature. Meaning, the models could not estimate the past temperature data without allowing for a very strong impact of greenhouse gases on temperature.

Thus, the climate models provide further evidence for the influence of human-emitted greenhouse gases on earth's temperature.

Effects

Several environmental impacts are associated with human-caused climate change[44]. The climate monitoring tools discussed earlier have provided an enhanced understanding about these effects.

- Warming oceans: Over 90% of the excess heat from the greenhouse gases is taken up by the oceans[45]. This increases the ocean surface temperatures. The impact on the oceans is important because there is a two-way relationship between the oceans and the climate. Oceans can impact climate, just as climate can impact oceans. For example, warmer waters can promote the development of more powerful storms.
- Shrinking icesheets, retreating glaciers, decreasing snow cover, and declining artic sea ice: The general trend shows a significant increase in these effects over the last few decades.
- Rising sea levels: The combined effect of thermal expansion from the warming of oceans and melting of ice is causing a rise in the sea levels. The global sea levels have risen by about 0.2 meters (8 inches) over the last century. The sea levels have increased faster over the last two decades compared to the last century.
- Acidification of oceans: Oceans have absorbed 20 to 30% of the human-caused CO_2 emissions in recent decades and have been acidifying because of this uptake of extra CO_2.
- Greening effect: The elevated CO_2 concentration has enhanced global photosynthesis., i.e., it has had a fertilization effect on plants[46].
- Increasing severe weather events: Extreme events are on the rise globally since the past several decades.

The Intergovernmental Panel on Climate Change (IPCC) was set up to advise the global governments about critical issues related to climate change. The importance of human-caused climate change can be recognized from this summary statement from one of its flagship reports[47]:

"*Continued emission of greenhouse gases will cause further warming and long-lasting changes in all components of the climate system, increasing the likelihood of severe, pervasive and irreversible impacts for people and ecosystems. Limiting climate change would require substantial and sustained reductions in greenhouse gas emissions which, together with adaptation, can limit climate change risks.*"

...§§§-§-§§§...

2. History of Climate Change

"If I have seen further, it is by standing on the shoulders of giants."–
Sir Isaac Newton

Local and national governments around the globe are emphasizing efforts related to climate change mitigation. The emphasis is a result of the appreciation of climate change science and international collaborative efforts. A discussion about the history of climate change science and international efforts is instructional for understanding how we got here.

History of Climate Change Science

The climate system consists of five components: atmosphere, biosphere, cryosphere, hydrosphere, and land surface[48]. These components intricately interact with each other (**Figure 2.1**). Because of the complexity, the current understanding of climate change science has only been possible because of extensive efforts. This section provides a timeline of the critically important efforts.

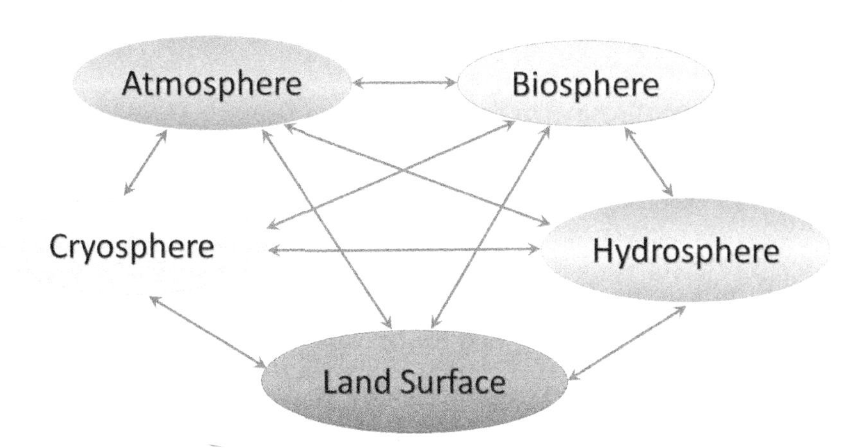

Figure 2.1: Components of earth's climate system and their interactions.

Warming effect of greenhouse gases

The French mathematician Joseph Fourier proposed that air could trap heat energy back in 1824[49]. The Irish physicist John Tyndall made a significant advance a few decades later[50]. He demonstrated that certain components of the air such as water vapor and CO_2 had special properties. Specifically, he showed that the water vapor and CO_2 molecules had the strong ability to absorb heat energy. He attributed this to the higher complexity of these molecules compared to the simpler nitrogen and oxygen molecules.

Svante Arrhenius, the Swedish Nobel Laureate, studied this phenomenon to understand the ice age cycles. At the end of the 19th century, he proposed that a substantial change in CO_2 levels would impact glacial advances and retreats. He also speculated that CO_2 emitted from coal combustion would play a significant role in defining the earth's temperature in the future centuries[51]. For decades, the scientific community ignored Arrhenius' efforts on this topic.

The English Engineer Guy Callendar revived the greenhouse gas warming theory in the 1930s[52]. He proposed that warming was already underway from the CO_2 produced from fossil fuels.

The warming effect from other greenhouse gases was identified relatively recently. The NASA scientist V. Ramanathan discovered the warming effect of fluorinated gases (CFCs) in 1975[53]. The warming effects of atmospheric methane and nitrous oxide were identified the following year[54].

Climate monitoring

Tools for measuring temperature, atmospheric pressure, wind speed, wind direction and rainfall were available by the end of 17th century. The accuracy of the measurements was improved over the next two centuries. This was accomplished by developing superior instruments and methods[55].

A very important development in the field occurred in the second half of the 20th century. The first dedicated weather satellite[56], TIROS 1, was placed in the orbit on April 1, 1960 (**Figure 2.2**).

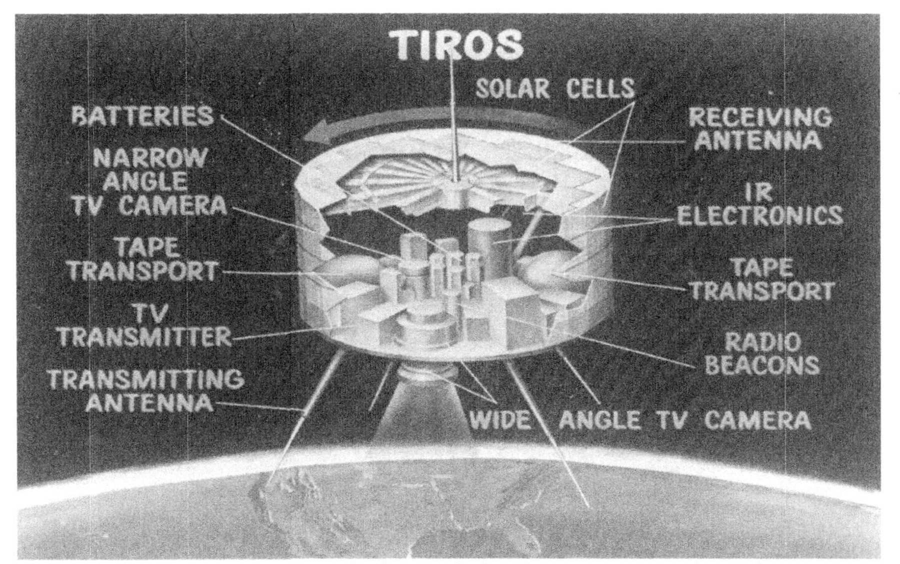

Figure 2.2: An artist rendering of the instruments aboard TIROS 1. Source: NASA[57].

Rapid advances made satellite technology a powerful tool for climate monitoring. Satellites have provided a wealth of data on atmospheric temperature, humidity, global sea surface temperature, snow cover, sea and land ice, global vegetation, volcanic eruption, activity of the sun, sea level rise, and more[58]. Recall, satellite technology played a key role in understanding the influence of solar variation on climate change[59].

The American scientist Charles Keeling made a very important contribution in the middle of the 20th century[60]. He initiated the high accuracy measurements of atmospheric CO_2. These measurements served as the starting point for studying the impact of burning fossil fuels on the atmospheric CO_2 concentration. This work also started the trend for accurate monitoring and record keeping of the other greenhouse gases.

The data from these climate monitoring tools have led to a robust understanding of the climate. For example, it led to the development of the global temperature time-series (Figure 1.1), which established the connection between the rising atmospheric CO_2 and global surface temperature[61]. For reference, the global temperature time-series includes several hundred million individual data points collected over a century.

Climate of the distant past

Understanding the climate of the distant past has been a major scientific goal over the last century.

The Serbian mathematician Milankovitch made a critical contribution in the first half of the 20th century[62]. He postulated that the earth's orbital changes were responsible for activating the start and end of the ice ages.

Advances in paleoclimatology allowed the validation of Milankovitch' theory. A key advance was the discovery that the sediments obtained from deep in the sea could be used to study past climates. In the 1970s, the deep-sea sediment core method was used to generate a time-series of climate data which extended over a period of several hundred thousand years[63]. The study showed that the orbital changes played a role in the ice age cycles. Thus, it provided empirical support for Milankovitch' theory.

Other methods for studying the past climate–such as tree rings, corals and ice cores–were developed in the 1960s, 1970s and 1980s, respectively[64]. These methods have played a major role in understanding climate change. Most importantly, scientists were able to distinguish between natural and human-caused climate change because of these methods.

Understanding the severity of human-caused climate change

Arrhenius in the early 1900s and Callendar in the mid-1900s played a pioneering role in connecting human emitted greenhouse gases to the warming of earth. Both believed that this human-caused warming would be beneficial[65,66]. This shows that the scientific community had no concern about the CO_2 impact until the middle of the nineteenth century.

Over the next decades, climate scientists recognized the risk potential from increasing atmospheric CO_2. But the scientific community was not overly concerned until the mid-1970s. Two pieces of information attest to this[67].

First, the premier scientific committee formed in 1965–for advising the U.S. President about environmental issues–did not recommend any specific action to control CO_2 emissions[68]. Although they discussed the possibility of significant future risk, their only recommendation was to continue the precise monitoring of CO_2 and temperature in the stratosphere. They made no recommendation about reducing CO_2 emissions.

Second, a high-profile U.S. National Science Academy report issued in 1975 acknowledged that the scientific community had a poor understanding about climate change[69]. The report issued several recommendations to improve the understanding.

Subsequent research raised the level of concern in the scientific community, which led to the formation of IPCC. But a definitive understanding was missing even as recent as the early 1990s.

A definitive understanding requires unequivocal detection, i.e., conclusive evidence. The first flagship IPCC report published in the early 1990s acknowledged that a definitive understanding of the severity of human-caused climate change was lacking at the time of the report. This is evident from the statements in the judgment section of the report:

- "*The size of the warming is broadly consistent with predictions of climate models, but it is also of the same magnitude as natural climate variability.*"
- "*The unequivocal detection of the enhanced greenhouse effect from observations is not likely for a decade or more.*"

As predicted in the judgment section, the scientific community definitively understood the severity of human-caused climate only towards the end of the twentieth century[70,71].

Thus, contrary to widely circulated myths, the scientific community has only had conclusive evidence about the severity of human-caused climate impact since the past few decades.

International Efforts

Pollutants typically cause local environmental problems. For example, if an industrial process in a Chinese city releases a soil pollutant, its environmental impact will be felt locally. The city administration will be able to address the problem by regulating the offending industrial process.

The situation is different for the environmental issues from greenhouse gases. The greenhouse gas concentration in the atmosphere is practically the same at all geographical locations[72]. It does not depend on the location of the emissions. For example, the greenhouse gases emitted by a city in North America will impact countries all over the globe. This is because the greenhouse gases have sufficient lifetime to spread uniformly around thc globe[73].

Thus, greenhouse gas emissions are a global problem. Consequently, global collaborative efforts are essential for addressing human-caused climate change. Three major international efforts are discussed below.

Intergovernmental panel on climate change (IPCC)

IPCC was established in 1988 by the World Meteorological Organization and United Nations Environment Programme[74]. It currently has 195 member countries.

The primary directive of IPCC is to inform the global Governments about the key aspects related to climate change based on the collective global research. Key aspects include the scientific basis for climate change, impacts, future risks and options for adaptation and mitigation.

IPCC relays such information to the governments and the general population via flagship reports that are published every few years. These reports include recent research from the scientific community. Specifically, each report contains timely information collected from thousands of studies undertaken by tens-of-thousands of researchers.

IPCC released its first flagship report in 1990, commonly known as the First Assessment Report or FAR[75]. Subsequently, it released four more synthesis reports: 1996 (Second Assessment Report, SAR), 2001 (Third Assessment Report, TAR), 2007 (Fourth Assessment Report, AR4) and 2014 (Fifth Assessment Report, AR5)[76]. The next synthesis report (Sixth Assessment Report, AR6) is due for release shortly[77].

Member governments and participating organizations provide input for selecting the authors and reviewers for these reports. The reports undergo a rigorous review by experts and governments[78]. For example, more than 800 scientists from over 80 countries developed the 2014 IPCC report[79]. Over 30,000 scientific papers were assessed. The report was reviewed by over 2000 reviewers, who provided around 140,000 comments.

Kyoto Protocol

The Kyoto protocol was an international treaty aimed at reducing greenhouse gas emissions. It was adopted in 1997 and entered into force in February 2005[80]. It set binding greenhouse gas emission reduction targets for industrialized countries and transitioning economies. Overall, the target was 5% emission reductions compared to 1990 levels over the 2008-2012 commitment period.

The protocol did not restrict emissions from countries such as China and India because of their economic status. The United States did not ratify the Kyoto protocol. Therefore, it was also not bound by the Kyoto protocol.

The protocol was a significant diplomatic accomplishment at the time of its adoption. However, its effectiveness was questioned in the following years because of its narrow participation and modest targets. The commitment was extended for a period from 2013 to 2020 via the Doha Amendment in 2012. However, the Doha Amendment never entered into force.

The Kyoto protocol has been effectively replaced by the Paris Agreement.

Paris Agreement

The Paris Agreement was adopted on December 12, 2015, by 197 countries and went into force on November 4, 2016. It is a landmark achievement because it brought countries around the globe together for the first time to combat climate change via a binding agreement[81].

A primary goal of the Paris Agreement is to limit earth's warming to well below 2° C, preferably to 1.5^{o} C, compared to preindustrial levels.

To achieve the goal, each member country is expected to combat climate change via actions that are based on its nationally determined contribution (NDC). For reference, the United States in its first NDC agreed to reduce its total greenhouse gas emissions by 26-28% below the 2005 emission level by the year 2025[82].

Countries have significant flexibility in stating their NDC. For example, China has connected its CO_2 reductions to its gross domestic product (GDP). Specifically, it has pledged to lower its CO_2 emissions per unit of GDP by 60-65% in 2030 from its 2005 emission level[83]. GDP is the monetary value of all goods and services made within a country during a specific period. China's GDP has been growing faster than its CO_2 emissions. This means that it can meet its pledge and yet have substantially higher CO_2 emissions in 2030 compared to 2005. For reference, China's historical CO_2 emissions and CO_2/GDP emissions are shown in **Figure 2.3**.

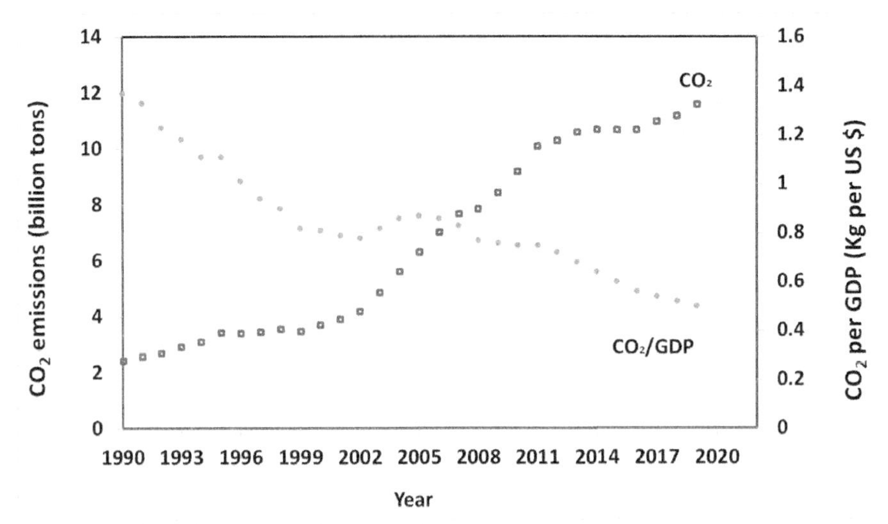

Figure 2.3: Historical data on CO_2 emissions and CO_2 emissions per unit of GDP for China. Source: Netherlands Environmental Assessment Energy[84].

Countries are expected to state their NDCs based on their history of greenhouse gas emissions, state of economy and technology availability[85]. For example, countries that have emitted more greenhouse gases are expected to share a higher responsibility for decreasing emissions.

Over 190 countries have ratified the agreement[86]. Only a few countries–Eritrea, Libya, Iran, and Yemen–have never formally accepted the Paris Agreement. Iraq and Turkey recently ratified the agreement. The United States has been an unusual case because of the change in direction based on change in administrations. The Obama administration issued an acceptance of the agreement on September 3, 2016. The Trump administration issued a withdrawal from the agreement on November 3, 2019. Following his election victory, President Joe Biden signed an Executive order on his first day in office to readmit the United States in the Paris Agreement.

...§§§-§-§§§...

3. Climate Change Debate

*"Nothing in life is to be feared, it is only to be understood. Now is the time to understand more, so that we may fear less"–**Marie Curie***

Human-caused climate change is a hotly debated topic. Two questions are critical to the debate. Is human-caused climate change a serious problem? And why the significant misunderstanding?

Herein, facts are presented, and implications are discussed to address these critical questions.

Is Human-Caused Climate Change a Serious Problem?

Two separate questions must be considered to address this question.

What is the conclusion from the scientific efforts?

Facts: Thousands of climate scientists have gathered vast amounts of climate data using a suite of climate monitoring tools[87,88,89]. The scientists have used the climate data and established scientific principles to understand the climate system. These studies have been documented in over a hundred thousand scientific papers[90]. Based on these studies, most climate scientists believe that human-caused climate change is a serious problem[91,92].

Discussion: Earth's climate system is extremely complex. Hence, adequate expertise is required to robustly understand this topic. Most climate experts believe that human-caused climate change is a serious problem.

The general population has a choice to either believe most of the climate experts or place trust in the few outliers.

Climate experts have presented evidence that strongly supports the conclusion that human-caused climate change is a serious problem[93]. For reference, six comprehensive IPCC reports have been published that summarize the evidence from tens-of-thousands of research papers[94].

Major scientific organizations, which represent all scientific fields, also support the conclusion[95].

For further facilitating the discussion, we will consider a hypothetical example. Assume that a loved family member has intense pain in the vicinity of the heart region. Diagnosis is sought from ten heart specialists–all with decades of relevant experience. Data is obtained from a series of medical tests. Nine heart specialists provide a diagnosis of a serious heart defect and recommend surgical treatment as soon as possible. However, one heart specialist indicates a low cause for concern and suggests a wait-and-watch approach.

How would you decide on a path forward for your loved one?

- Would you trust the common conclusion from the nine heart specialists? Or,
- Would you trust the one heart specialist with a different conclusion? Or,
- Would you develop expertise on heart diseases by doing internet research to decide the path forward?

Most will choose the path forward that is statistically more likely– i.e., the path forward that is based on the common conclusion from most of the experts. The alternatives are significantly less likely because of lesser or inadequate experience. Similarly, we can either believe most of the climate experts or place our trust in the unlikely alternatives.

Is the scientific community colluding?

Facts: The scientific community is extremely diverse[96]. It is not controlled by any one country or institution or any group of individuals.

Discussion: Different countries have different agendas. Most climate researchers from different countries agree that climate change is a serious problem. Furthermore, the conclusion is accepted by the scientific community. Major scientific organizations across the globe support the conclusion that human-caused climate change is a serious problem[97-104].

For reference, a list of roughly two hundred such organizations is provided in Appendix 1[105]. Such level of agreement amongst an extremely diverse population suggests with high confidence that the conclusion is driven by science.

Why the Significant Misunderstanding?

The general population is at-least somewhat concerned about the environment. The population can be classified into two groups based on the level of concern.

The first group considers environmental issues as their topmost priority. This group needs minimal convincing about human-caused climate change.

The second group cares about the environment but is also concerned about other issues such as economy, national debt, and energy security. This group needs systematic convincing about the seriousness of human-caused climate change. Why? Because this group is concerned that an incorrect diagnosis about climate change would detrimentally impact the other important issues. Consequently, this group requires consistent messaging about climate issues from credible sources. Unfortunately, this has not happened.

Inappropriate messengers and ineffective messaging are two major reasons that have contributed to the significant misunderstanding in the second group.

Inappropriate messengers: Polarizing politicians

Fact: Most countries are highly divided along political lines. For examples, polls in the United States have shown that the partisan divide between Republicans and Democrats has increased drastically over the past decades[106,107]. Many Democrats have a deep distrust of Republican political figures and vice versa.

Discussion: When the political divides are extreme, it is critical to avoid the perception of partisanship for environmental issues. This is unlikely if the messengers are deeply polarizing. Consequently, climate-related messaging by high-profile Democrats has inadvertently created mistrust about climate change amongst many Republicans[108-111]. This is a major concern because a large fraction of the Republicans belongs to the second group.

Such messengers have accelerated the understanding about human-caused climate change within the first group. However, accelerating the first group has a limited value because that group is easily convinced. Overall, these deeply polarizing messengers have caused a setback by disengaging the second group.

On the Republican side, certain high-profile politicians have increased the misunderstanding by marginalizing the importance of climate change[112,113].

Inappropriate messengers: Celebrities

Fact: Most movie stars, music artists, sports stars, and billionaires have an ultra-lavish lifestyle. Their lifestyle includes mansions, superyachts, personal jets, etc. Such a lifestyle leads to greenhouse gas emissions that are a hundred to thousand times more than an average person[114-117].

Discussion: When such celebrities preach that human-caused climate change is an extremely serious problem, there is a distinct lack of credibility. This hypocrisy has increased mistrust in the second group[118,119].

Ineffective messaging

Fact: The messaging from certain climate activists has been misleading. It has been substantially more pessimistic than the scientific consensus[120,121,122]. Examples include statements such as "*the world is going to end in 12 years if we do not address climate change*".

Discussion: Certain climate activists seem to suggest that the planet is currently on the brink of destruction because of human-caused climate change. Such exaggerated messaging is ineffective for achieving sustainable progress because it creates mistrust in the second group. It is more important to on-board the second group than to increase excitement in the first group. A powerful approach to convince the second group is by consistently providing believable information. This can be achieved by avoiding exaggerations and misleading information.

It is admirable that climate activists are extremely passionate about climate change. However, their emotional pleas can leave the perception that climate change is the only important issue. Many consider such pleas to be misplaced considering that the global society has been very slow to address many life and death issues–such as millions of preventable child deaths, hundreds of millions suffering from hunger and lack of access to basic healthcare[123,124,125].

Each day is a struggle for existence for hundreds-of-millions of people living in poverty. Clearly, poverty is currently a far more urgent problem. Incidentally, poverty is not limited to developing nations. United States, one of the most powerful economies, has around 37 million people living in poverty[126].

An often-repeated message is that climate change should be addressed with the highest urgency for the sake of the poor. This type of messaging is misleading because the impact from climate change is a secondary issue for the economically unfortunate. Poverty is their primary issue. The impact from climate change is much worse for the poor because they have limited access to energy and healthcare. If the goal was to urgently help the poor, an effective approach would be to urgently increase access to healthcare, energy and opportunities. Decreasing poverty would be far quicker than mitigating climate change[127,128].

Also, climate change adaptation is more important for the poor from an urgency perspective. Climate change adaptation involves anticipation of the adverse effects of climate change and taking appropriate action to minimize damage[129]. Therefore, benefits of climate change adaptation will be felt very quickly by the poor. In contrast, decades will be required for meaningful benefits from climate change mitigation. For example, if climate change is causing floods in a poor community, the urgent need is to build a flood levee to protect the community against flooding. Climate change mitigation efforts will not decrease the flooding impact in a meaningful timeframe. Hence, suggesting that climate change mitigation should have the highest priority for the sake of the poor is misleading[130]. Such messaging has contributed to credibility challenges around climate change.

Overall, ineffective messaging and inappropriate messengers have created credibility issues about the seriousness of climate impact. This has made the second group more susceptible to the misinformation that is continually being spread in different forms of media.

It is essential to onboard most of the global population to successfully address the problem. Alienating a significant fraction will disallow sustainable solutions. A successful path forward requires credible messengers and effective messaging. This major gap in climate efforts has caused major misunderstanding.

...§§§-§-§§§...

4. A Brief History of Energy

*"Energy availability is the pillar for social and economic progress in a society"–**United Nations Department of Economic and Social Affairs***

The dependence of the global society on energy cannot be overstated. Consequently, an energy transition must occur efficiently, i.e., in a manner that does not disrupt the energy supply.

Decisions for an efficient transition require an objective analysis of all crucial aspects related to energy. A key step is to understand how energy evolved over time, i.e., the history of energy.

Two periods define the history of energy: energy use before fossil fuels, and the fossil fuels era. The transition to low carbon energy is just getting started and will be discussed in later chapters.

Energy Use before Fossil Fuels

Early humans used biomass because it was accessible and easy to use. Twigs and branches could be collected and physically carried. The biomass could be burned to provide heat and cook food.

Biomass was an energy source of choice because it was plentifully available and cheap. Improvements in gathering and transportation methods ensured the availability and low cost of biomass for a long time[131]. Consequently, biomass remained the dominant energy source until the end of the nineteenth century[132].

Wind and water were used as energy sources to grind grain, cut wood and more. Wind energy was harnessed via windmills, while water energy was harnessed via water wheels. Fires, torches and whale oil were used for illumination.

Transportation of people and goods was based on muscle power (e.g., horses), waterpower (e.g., flowing streams), and wind power (e.g., sail boats). As with biomass, most of these energy sources were used for centuries[133]. Technology advances were incremental during this time.

Fossil Fuels Era

Coal, natural gas and crude oil are termed as fossil fuels. They were formed from the remains of dead organisms[134]. Dirt and water covered the organisms after their death. Heat and pressure applied over millions of years converted the dead organisms to fossil fuels. Coal was formed from dead plants while natural gas and crude oil was formed from dead marine organisms.

Fossil fuels have huge reserves because they were formed from plants and marine organisms that were abundant all over earth.

The three fossil fuels are substantially different in terms of their physical and chemical characteristics. Coal is a solid, crude oil a liquid and natural gas occurs in the gaseous form. The specific characteristics of the fossil fuels define their suitability for different energy applications. Thus, the different fossil fuels need to be discussed separately.

Transition to coal

The transition to fossil fuels began with coal. In England, the decreasing supply of local wood prompted the shift in the sixteenth century. Local forests were gradually disappearing because of the increasing demand for wood[135]. The cost was high for transporting wood from distant locations to populated centers.

Coal presented an excellent alternative because of its abundance and low cost. Also, coal was attractive because of its high energy content and concentrated deposits. On the flip side, coal mining was an inefficient process.

An Englishman, Thomas Newcomen, addressed the inefficiency problem. He designed a practical steam engine, which increased the efficiency of mining coal[136].

More importantly, the steam engine provided an efficient method to convert heat to motion. This unlocked several new applications of heat. The increase in mining efficiency coupled with new applications led to a large increase in coal use. The Scottish inventor, James Watt, improved the design of the steam engine–which further accelerated the use of coal.

Design of chimneys, fireplaces and flues were improved, which decreased the concerns about the foul smell and smoke from burning coal[137].

Such advances helped the growth of coal use. Coal's share as an energy source in England increased from 10% in the mid-sixteenth century to over 90% in the mid-nineteenth century[138].

The United States had large availability of wood and animal fuel[139]. But these energy resources gradually became constrained because of the high consumption caused by the increasing population and economic growth. Coal was considered as an alternative energy source because of its massive reserves.

Over time, the price of coal dropped significantly because of the expansion of coal mining and railroad construction. For reference, the coal price decreased by a factor of three from the 1830s to the 1860s[140]. The low cost and increasing need for energy accelerated the use of coal in the United States. Coal became the dominant source of energy in the United States by the end of the nineteenth century[141].

Globally, coal overtook traditional biomass and became the dominant source of energy in the early twentieth century[142]. Thus, it took coal over two centuries to replace biomass as the dominant source of energy.

Currently, coal contributes to 28% of the total energy consumed globally (**Figure 4.1**)[143,144].

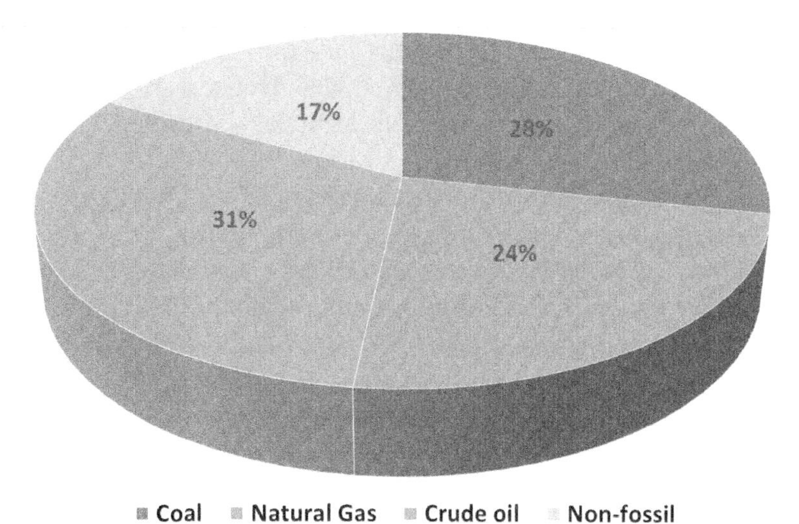

Figure 4.1: Average contribution from different energy sources to the total global energy consumed in recent years[145]. Source: U.S. Energy Information Administration.

Transition to liquid petroleum fuels (Crude Oil)

The transition to crude oil occurred after coal[146]. It was driven by transportation applications.

Horse carriages were used for personal transportation until the early part of the twentieth century. In 1900, for example, London had over 50,000 horses for city transportation and New York had around 100,000 horses[147].

The horses consumed large quantities of food and produced proportionate amounts of manure. Millions of pounds of manure were produced daily, which caused a waste disposal problem. Also, the horses had a short life expectancy for transportation work. These problems paved the way for the transition to motorized personal vehicles.

The coal-fueled steam engine was used for powering trains and ships in the nineteenth century. However, this engine was not convenient for use in personal vehicles because of its large size. A much smaller engine was desirable for personal vehicles.

German inventors–Nicolaus Otto, Karl Benz, Gottlieb Daimler and Rudolph Diesel–played pioneering roles in the invention and development of the internal combustion engine. They undertook this work in the late nineteenth century[148]. The internal combustion engine was powered by liquid petroleum products produced from crude oil.

The small, yet powerful engine was the key to the appealing performance of motorized personal vehicles. Henry Ford, an American businessman, made the motorized vehicles affordable by mass producing them in early twentieth century.

Liquid petroleum fuels offered several advantages over coal such as higher energy content, easier on-board storage, and lower pollution[149]. Consequently, these fuels were found to be more suitable for trains, ships, and airplanes. This led to a growing demand. Several refineries were built for converting crude oil to petroleum fuels to satisfy the demand.

Crude oil products dominated the transportation sector by the middle of the twentieth century. Currently, crude oil contributes to 31% of the total energy consumed globally[150,151].

Transition to natural gas

Transporting gaseous fuel from the production site to the end users was challenging. Hence, the transition to natural gas occurred after the transition to coal and crude oil.

Natural gas was mainly used for lighting purposes in the nineteenth century[152]. The discovery of the Bunsen burner in 1885 prompted the heating applications of natural gas[153].

However, only a few natural gas pipelines were built before the mid-twentieth century because of technology limitations. This constrained the growth of natural gas.

The pipeline technology–types of metals used, welding technique, and pipe manufacturing methods–improved significantly in the 1940s[154]. This led to development of vast pipeline systems. For example, hundreds-of-thousands of miles of pipeline were built in the United States during the 1950s and 1960s[155].

The development of robust pipeline technology increased consumer access to natural gas around the globe. This increased accessibility allowed the use of natural gas for heating and cooking in homes and as a fuel in industrial plants. Congruently, the global consumption of natural gas increased five-fold from 1940 to 1960[156].

Electricity generation was the next big application of natural gas. Until the 1970s, coal was the feedstock-of-choice for generating electricity. But pollution from coal power plants caused substantial concern, especially in high population centers.

Natural gas power plants did not pollute like coal power plants. Also, several advances were made in natural gas power plant technology during that timeframe. Consequently, natural gas power plants became more attractive for electricity generation.

Since the 1980s, natural gas power plants have continued to increase in favorability. This was accelerated by the development of fracking technology. The new technology drastically decreased the price of natural gas. In turn, this lowered the electricity production costs in countries that had convenient access to natural gas, for e.g., United States[157,158].

In 2016, natural gas overtook coal for electricity production in the United States [159]. But coal continues to dominate in countries such as China and India, which do not have access to cheap natural gas.

Currently, natural gas contributes to 24% of the total energy consumed globally[160,161].

Key Lessons from Prior Energy Transitions

Superior cost and convenience of the new over the incumbent has driven prior energy transitions. In other words, the incumbent has been replaced only after the new energy resource or technology has achieved a cost and convenience advantage.

The cost of energy has been critical because of its financial impact on the society. The financial impact is a result of direct and indirect energy costs. Indirect costs are energy costs embedded in goods and services. For example, indirect costs include the energy costs incurred during the production and transportation of goods[162]. Thus, high energy costs have a large financial impact on the society.

Convenience of energy refers to the overall experience related to energy use. This characteristic has been important because energy is intricately woven into every aspect of the human society.

Throughout history, humans have selected the best energy resource or technology available in terms of cost and convenience. It has been challenging for a new energy technology to establish a cost and convenience superiority over an incumbent. Consequently, many decades to a few centuries have been required for the wide-scale deployment of new energy technologies. For reference, a technology is said to be deployed on a wide-scale when it becomes a major component in the energy mix.

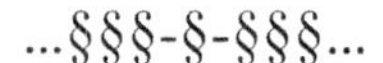

5. Introduction to Low-Carbon Energy

The energy sector generates roughly 75% of the global greenhouse gas emissions[163,164]. Hence, addressing the emissions from this sector is vital for climate change mitigation.

Low-carbon technologies emit drastically lower greenhouse gases than fossil fuel technologies over their life cycle. A lifecycle extends from the cradle to the grave. For example, the life cycle of a power plant includes extraction of raw materials, processing of materials, production of various components, integration of components, installation, power generation, maintenance, operation, decommissioning and disposal.

Different low-carbon energy options have been proposed for decreasing greenhouse gases. Many are included in the renewable energy category. Renewable energy is derived from natural resources that can be easily replenished. Examples of low-carbon options that do not belong to the renewables category include nuclear power and fossil fuel power with carbon capture.

The low-carbon energy options differ widely in terms of their key characteristics. Understanding their key characteristics is critical for understanding the efficacy of the options.

Key Characteristics

Key characteristics are defined as those that are crucial for apt selection between the different options. In subsequent chapters, these key characteristics are used to discuss the advantages and challenges of the different solutions.

The lessons from the previous transitions have been used to identify the key characteristics.

Key Characteristic 1: Cost

The cost characteristic has two components: upfront cost and total lifetime cost.

Upfront cost is the initial investment required to deploy the solution. For example, the cost to build a power plant or purchase a vehicle[165]. A low upfront cost ensures a high efficiency of greenhouse gas reduction by lowering the annual budget and decreasing financial risk.

Total lifetime or levelized cost is the cost associated with the entire operating life of the option. It includes all costs incurred over the lifetime such as the upfront cost, operating and maintenance costs and financing costs. A low total lifetime cost ensures a high overall cost efficiency of greenhouse gas reduction.

Key Characteristic 2: Convenience

Convenience refers to the availability and ease of use. The incumbents–fossil fuel technologies–meet the energy demand 24X7. Moreover, the use of energy from these technologies is effortless.

A similar convenience will be expected from the replacement low-carbon options. If a low-carbon option is not self-sufficient in providing the required convenience, a supporting technology will be needed.

Approach for Discussing the Low-Carbon Energy Options

The low-carbon options are compared in terms of their relative performance for cost and convenience in the next two chapters.

Each option is designated as either low, midrange, or high cost depending on its relative cost performance within a given sector. A top quartile performer is designated as a low-cost option, while a bottom quartile performer as a high-cost option.

The cost discussion is based on data from government and intergovernmental sources. Average global costs are discussed unless otherwise noted.

For the convenience characteristic, the options are designated as being either advantaged or challenged. For example, if an option provides a similar level of convenience as fossil fuel technologies, it is characterized as being advantaged. But if the low-carbon option requires a supporting technology for providing the required convenience, it is characterized as being challenged.

Other key issues are also discussed. For example, the severity level of environmental impacts for the different technologies is included[166]. Environmental impacts include health and safety impacts.

Overall, the discussions are designed to provide a robust understanding about the wide-scale deployment of low-carbon energy options.

The options have been divided into two groups: low-carbon power options and low-carbon transportation options. The highlights are discussed in the next two chapters. Over the years, a barrage of media reports has led to serious misunderstandings about these options. These are discussed in a separate chapter.

...§§§-§-§§§...

6. Low-Carbon Power Options

"We will make electric light so cheap that only the rich will burn candles."–***Thomas Edison***

The electric power sector is the largest direct emitter of CO_2. It contributes to 40% of the greenhouse gas emissions from the global energy sector[167,168].

The life cycle greenhouse gas emissions from low-carbon power technologies are small compared to fossil fuel technologies (**Figure 6.1**). This has led to an interest in the use of electrification for reducing greenhouse gases. Electrification is the process of replacing technologies that use fossil fuels with technologies that use electricity[169].

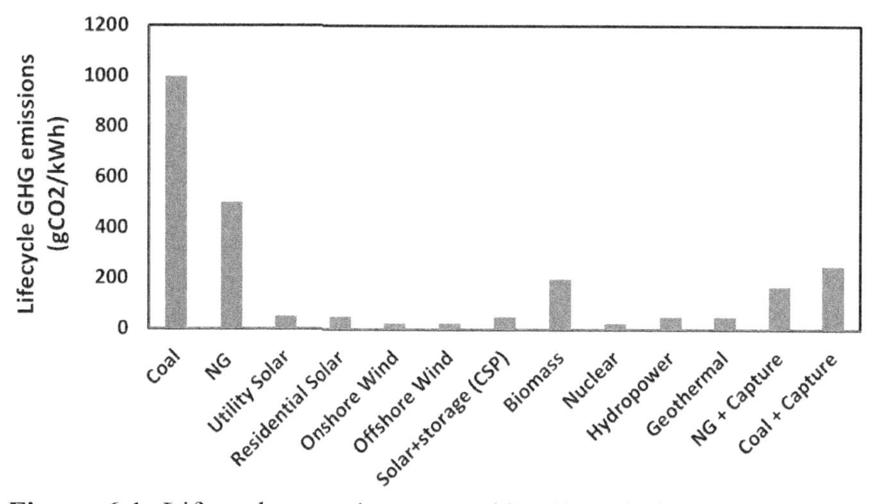

Figure 6.1: Lifecycle greenhouse gas (GHG) emissions from different electricity generation technologies[170, 171,172,173]. NG = natural gas. CSP = concentrated solar-thermal power. Capture = CO_2 capture and storage.

The United States Energy Information Administration (U.S. EIA) classifies the technologies into two groups: dispatchable and resource-constrained[174,175]. In case of dispatchable technologies, operators can adjust the output to meet the 24X7 electricity demand. In contrast, the operators cannot control the output to meet the demand in case of resource-constrained technologies. The electricity generated from resource-constrained technologies is intermittent because it depends on weather conditions.

This lack of flexibility is a major disadvantage for resource-constrained technologies. Because of this critical distinction, the cost characteristics of dispatchable and resource-constrained technologies cannot be compared in a standalone fashion. Instead, the cost and convenience characteristics must be considered simultaneously.

Solar Power

Only a tiny fraction–about a billionth–of the energy radiated by the sun reaches earth. Yet it is enormous! The amount of solar radiation received by earth in an hour and a half is comparable to the total energy consumed by the global population in a year[176]. Thus, an extraordinarily large amount of solar radiation is theoretically available for generating electricity.

The first application of solar power was in the 1950s[177]. But solar power has been used in significant amounts only in recent decades.

Costs have declined rapidly, which has led to rapid growth in solar power[178].

Electricity generation from solar has increased by over a factor of five hundred over the past couple of decades[179]. Despite the impressive growth, the global share of solar power in the electricity generation sector is currently very low–less than 4%[180,181,182].

Environmental impacts are low for solar power compared to other options in the power sector[183,184,185].

Solar photovoltaic (PV) is the predominant solar technology. Hence, it is the focus of further discussions herein.

Technology Basics

Solar PV technology can directly convert solar radiation into electricity. The basic component of solar PV technology is a solar cell. It uses semiconductor materials to convert solar radiation into electricity. Semiconductors have higher conductivity than insulators but lower than metals.

Different types of semiconductor materials have been used–such as silicon, cadmium telluride thin film, copper indium gallium selenide thin film, organic polymer, and quantum dot[186].

The choice of the semiconductor material defines the cost and performance. Silicon-based technology has been most widely used because of its advantage over other materials in terms of cost, conversion efficiencies, and life[187].

Individual solar cells are small and produce 1 or 2 watts of power. To increase the output, the solar cells are connected to form modules or panels. The modules are individually used or further connected to increase the output.

The modular nature of solar PV technology allows a wide range of output–from small to very large. Hence, solar PV technology is suitable for small-scale residential as well as large-scale utility power applications.

Utility Solar: Advantages and Challenges

Utility solar has a low upfront cost and low total lifetime cost compared to other options in this sector[188-192].

For reference, the upfront and total lifetime costs for the different electricity generating technologies in the United States are provided in **Figures 6.2** and **6.3**[193-197]. For the power sector, total lifetime cost is commonly known as the levelized cost of electricity (LCOE).

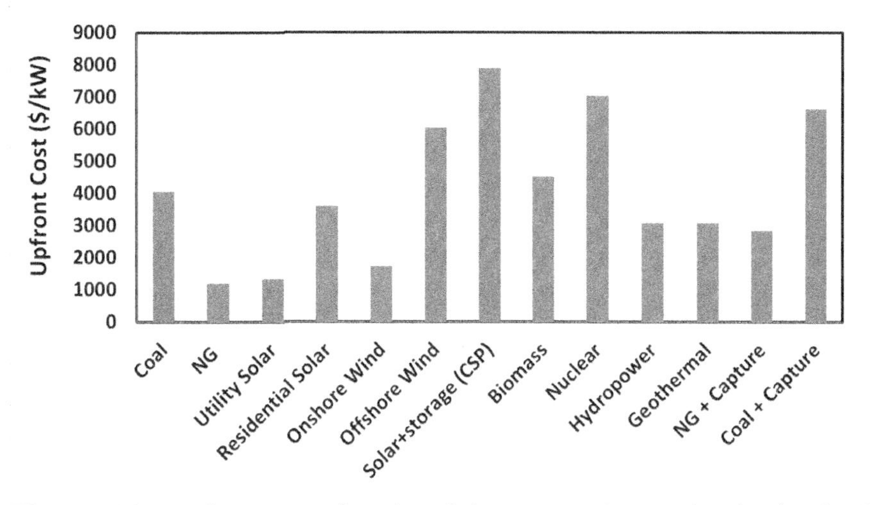

Figure 6.2: Upfront cost for electricity generation technologies in the United States. Costs are in 2021 $/kW. NG = natural gas. CSP = concentrated solar-thermal power[198]. Capture = CO_2 capture and storage. Primary Source: U.S. EIA

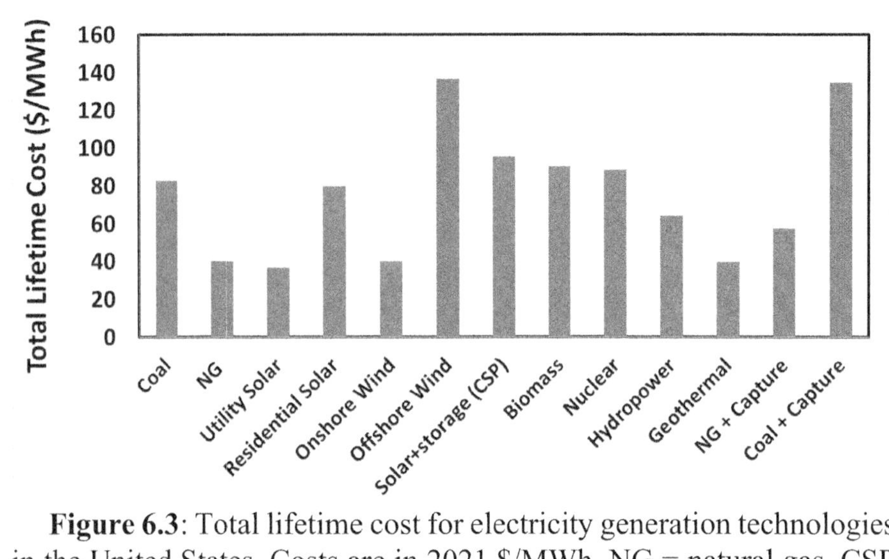

Figure 6.3: Total lifetime cost for electricity generation technologies in the United States. Costs are in 2021 $/MWh. NG = natural gas. CSP = concentrated solar-thermal power[199]. Capture = CO_2 capture and storage. Primary Source: U.S. EIA

Solar technologies suffer from a convenience or intermittency challenge. Sunlight has day-to-day and seasonal variation. Consequently, solar technology cannot supply 24X7 electricity on a standalone basis. Dispatchable technologies such as a natural gas, coal, and nuclear power do not suffer from this challenge.

Solar power is a low-cost option only if it is deployed to a limited extent. A brief explanation is provided below.

Solar power plants provide electricity only when there is sunlight. This limitation causes an imbalance in supply and demand. Consequently, the dispatchable power plants in the electrical grid must be adjusted up or down to balance supply and demand.

When the sun is shining, the electricity production from dispatchable power plants must be decreased to compensate for the production from solar power. In contrast, when the sun is not shining, the electricity production from dispatchable power plants must be increased. The adjustments depend on the amount of solar power included in the electrical grid. More the amount of solar power, greater is the adjustment required from the dispatchable power plants.

There are economical limitations to the extent to which the dispatchable power plants can be adjusted. Large suboptimization of the electrical grid is expected when solar power deployment exceeds a certain level[200]. Thus, only a limited deployment of stand-alone solar power is practical in any electrical grid.

There is a solution[201]. Solar power plants can be widely deployed if supplementary technologies such as energy storage are also widely deployed. Energy storage can circumvent the intermittency of solar power by storing electricity for later use when sunlight is not available. Related challenges are discussed later.

Utility solar power plants require over hundred times more land than natural gas power plants[202,203]. Hence, land availability can be a challenge for utility solar in highly populated regions.

Residential Solar: Advantages and Challenges

Residential solar has a midrange upfront cost and a midrange-to-high total lifetime cost compared to other options in this sector[204]. Residential solar has substantially higher upfront cost and total lifetime costs compared to utility solar[205,206,207]. For example, the total lifetime cost for residential solar is over two times that of utility solar. The cost disadvantage for residential solar is related to its poor economy of scale relative to utility solar. Economy of scale refers to the principle that a larger-scale production has lower costs per unit of production compared to a smaller-scale.

Residential solar is also inconvenienced by the intermittency challenge like utility solar. Likewise, its wide-scale deployment requires energy storage.

A key advantage of residential solar is that it can be installed on a roof top. Hence, it does not require large additional land unlike utility solar.

Solar Power with Energy Storage

Long duration energy storage is a solution to the intermittency problem of solar power. However, only a miniscule fraction of solar power is currently associated with long duration energy storage[208,209].

The electricity needs to be stored for long durations–many hours to a few days–to address the variation in sunlight availability. Several long duration energy storage technologies have been considered over the years[210]. Examples include thermal energy storage, pumped hydropower, compressed air energy storage and battery technologies. Pumped hydropower and compressed air energy storage have several advantages over other long duration energy storage systems[211].

However, these technologies suffer from limited availability because of their specific requirements related to water bodies or caverns[212,213]. This is a crucial obstacle to the wide-scale deployment of these technologies.

Therefore, energy storage technologies such as thermal storage systems, battery storage systems, and green hydrogen are more practical for wide deployment. Solar in conjunction with these three energy storage systems will be considered herein.

Solar with energy storage has low-to-moderate environmental impacts[214,215,216].

Technology Basics

Concentrated solar with thermal storage: It consists of two components: concentrated solar and thermal storage. The concentrated solar component uses a combination of mirrors and receivers to collect and concentrate sunlight to produce heat energy. The thermal storage component stores the heat. A turbine-generator system is used to covert the heat to electricity[217]. Mineral oils and molten salts are used as heat transfer fluids.

Solar power with battery storage: The electricity produced by solar PV is stored for later use by using battery technology. Different types of battery technologies are being considered such as lithium ion, redox flow, lead acid, sodium-sulfur, sodium metal halide, and zinc hybrid cathode[218]. A battery consists of anode and cathode electrodes and an electrolyte. The battery technologies differ in terms of the materials of the electrodes and electrolyte and the configuration of the battery system[219].

Green hydrogen using solar power: In the first step, the electricity generated by solar power is used to split water molecules to produce hydrogen. This hydrogen is called green hydrogen. In the second step, green hydrogen is used as a fuel to generate electricity. Thus, hydrogen is used for energy storage in this technology.

Advantages and Challenges

Based on a global average, sunlight is available for less than seven hours per day. For example, the sunniest year in Germany had an average sunshine of five and a half hours per day[220].

Moreover, sunlight is unevenly available throughout the year. For example, countries have higher sunlight hours in summer compared to winter. Also, several days have very little sunlight, e.g., rainy days. Thus, massive energy storage is required to balance electricity supply and demand when using standalone solar power.

When coupled with wind power, energy storage requirements decrease but are still substantial. For example, over 24 hours of energy storage will be required for most locations when wind and solar power are optimally deployed[221].

Energy storage technologies are resource intensive[222]. For example, battery energy storage systems require large quantities of rare materials[223].

Because of these issues, solar power with energy storage has a high upfront cost and high total lifetime cost compared to other options[224,225,226,227]. The cost disadvantage is further heightened for residential solar with storage because of the poor economy of scale. Also, energy storage based on green hydrogen has very high upfront and lifetime cost because of poor energy efficiency.

Solar coupled with adequate energy storage can provide 24X7 electricity. Therefore, it does not suffer from an intermittency-related convenience challenge.

Wind Power

Wind is formed by a combination of three events: the sun unevenly heating the atmosphere, irregularities of the earth's surface and earth's rotation[228]. Theoretically, less than 30% of the available wind resources would be sufficient to supply the global energy needs[229]. Thus, wind energy has large potential.

The first wind turbine with significant capacity was installed in Cleveland, Ohio in 1888. But wind power has only recently gained prominence–driven mainly by climate change concerns. Costs have declined rapidly, which has accelerated growth in wind power.

The deployment of wind power has increased fifty-fold over the past couple of decades[230]. Despite this large growth, the global share of electricity produced by wind power is currently low, around 7%[231,232,233].

Wind power has low environmental impacts compared to other options in the power sector[234,235,236].

Technology Basics

Wind power technology harnesses the energy in the wind to produce electricity. Specifically, wind turbines collect the kinetic energy from wind using propeller-like blades. The energy in the wind rotates the blades around a rotor. The rotor is attached to a main shaft, which turns the generator to produce electricity. Most wind turbines have a propeller-style design with three blades that rotate around a horizontal axis[237].

The output from the wind turbine is proportional to the cube of wind speed[238]. Wind speeds increase with increasing distance from the ground. Wind turbines, therefore, need to be tall to access high wind speeds. Turbine towers have an average height of 90 meters, which is approximately the same height as the Statue of Liberty[239].

Wind turbines can vary substantially in capacity. Larger capacity wind turbines are more cost-effective because of the efficiencies from economy of scale. Utility-scale plants, which consist of a group of large wind turbines, are often described as wind farms. Onshore wind farms are land-based while offshore wind farms are located on large bodies of water such as rivers and oceans.

Onshore Wind Power: Advantages and Challenges

Onshore wind has low upfront cost and low total lifetime cost compared to other options in the power sector[240,241,242,243].

Although solar and wind power are both intermittent sources, wind power has a significantly higher capacity factor[244]. Capacity factor is the availability of the electricity generator over a fixed period. **Figure 6.4** compares the capacity factors for different electricity generation technologies[245]. Dispatchable technologies have substantially higher capacity factors compared to resource-constrained technologies.

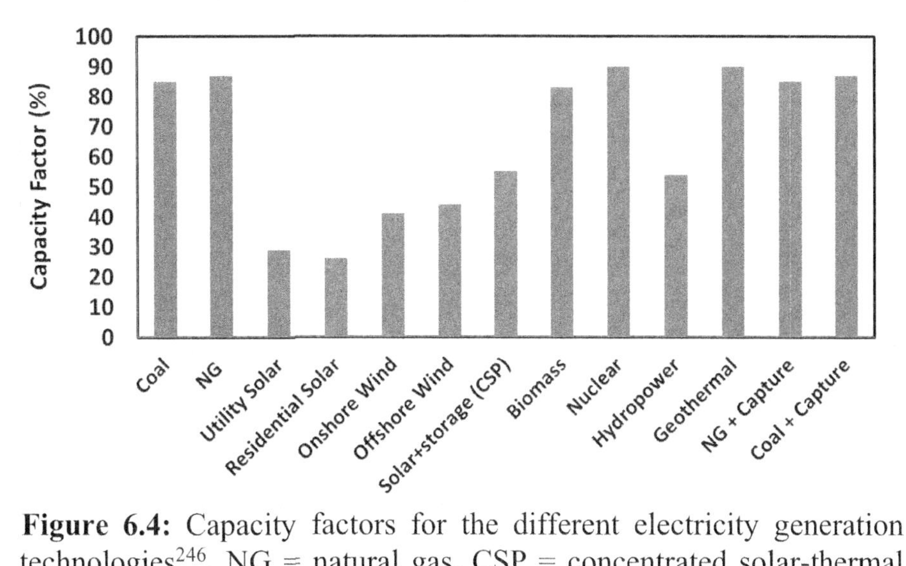

Figure 6.4: Capacity factors for the different electricity generation technologies[246]. NG = natural gas. CSP = concentrated solar-thermal power. Capture = CO_2 capture and storage. Primary Source: U.S. Energy Information Administration.

Wind power plants produce electricity when the wind is blowing. Because of this intermittency problem, it has a convenience challenge like solar power. Therefore, wide-scale deployment of onshore wind power requires a wide deployment of a supporting technology such as energy storage.

Onshore wind farms have very large land requirements[247]. For reference, they require several hundred times more land compared to natural gas power plants[248,249]. Therefore, land availability is a challenge for deploying onshore wind in highly populated regions.

Offshore Wind Power: Advantages and Challenges

Offshore wind has a midrange-to-high upfront cost and a midrange-to-high total lifetime cost compared to other options[250,251,252,253]. Offshore wind has markedly higher costs than onshore wind because offshore projects face more challenges related to installation and long-term robustness[254].

The wind resource is more abundant and uniform in case of offshore locations compared to onshore locations[255]. However, it remains an intermittent source. Therefore, offshore wind power also suffers from a convenience or intermittency challenge.

Offshore wind has a key advantage over onshore wind in that it does not require valuable land resources. This is important for locations where land availability is limited, for e.g., large urban centers.

Wind Power with Energy Storage

Wind power and solar power are comparable in terms of cost and convenience. Therefore, wind power with energy storage and solar with energy storage are also comparable, i.e., wind with adequate energy storage has high upfront cost and high total lifetime cost and does not suffer from an intermittency challenge[256,257].

Hydropower

Hydropower plants use the energy in water to produce electricity. Water moves through a vast global cycle which consists of evaporation from water bodies, cloud formation, precipitation, and accumulation of the precipitated water in the water bodies[258]. Hydropower is termed as renewable energy because the water cycle is a constantly recharging system.

The first large-scale hydropower plant was built in the 1880s in the United States. Many projects have since been deployed globally. Currently, hydropower is the dominant source of low carbon electricity and contributes to 16% of the global electricity[259]. From 1965 to 2020, hydropower produced nine times more electricity than solar and wind combined.

Hydropower has moderate environmental impacts compared to other options in the power sector[260,261,262].

Technology Basics

Hydropower plants convert the energy in flowing water into electricity. The two major types of hydropower plants include reservoir hydropower plants and run-of-the-river hydropower plants.

A reservoir hydropower plant consists of a dam to store water in a reservoir. When the water from the reservoir is released, it flows down from an elevation.

Run-of-the-river hydropower plants do not have reservoir. Instead, a diversion structure is used to access the energy in the flowing water.

For both, the energy in the flowing water drives a turbine, which in turn activates a generator to produce electricity[263]. The volume of the water flow and the change in elevation between the source and water's outflow determines the energy available in the flowing water. Higher water volume and a greater elevation change favors electricity production.

Advantages and Challenges

Hydropower has a midrange upfront cost and a low-to-midrange total lifetime cost compared to other options[264,265,266,267]. The costs are strongly influenced by the location and geological conditions[268]. A suitable location is especially important for hydropower plants because the location of the project determines the cost, environmental and societal impact.

Many hydropower projects have been completed over the past decades. Suitable locations are limited[269]. Therefore, availability of suitable locations for future projects is a major constraint.

Reservoir hydropower plants use the water stored in large reservoirs. This decreases the dependence on the variability of water inflows. Hence, reservoir hydropower plants provide significant flexibility to meet the electricity demand.

A similar flexibility does not exist for run-of-the-river hydropower plants, which depend on the water flow of the river. The water flow depends on the weather and thus, these plants are intermittent sources of electricity, i.e., they suffer from a convenience challenge. On the flip side, run-of-the-river projects are less complex and easier to implement.

Nuclear Power

Nuclear power is not considered as renewable energy because nuclear fuel is not naturally replenished.

The first commercial nuclear power plant was deployed in the 1950s. Nuclear energy is second only to hydropower in terms of low-carbon electricity generation. It currently contributes to 10% of the global electricity[270]. For reference, nuclear plants have generated six times more electricity than combined solar and wind since 1965 to 2020[271].

France has the highest share of nuclear power in the electricity mix. Nuclear power plants provide about 70% of the total electricity in France. The United States produces more electricity from nuclear power plants than any other country. But the share of nuclear power in the electricity mix in the United States is only 19%[272].

Nuclear power has low-to-moderate environmental impacts compared to other options in the power sector[273,274,275].

Technology Basics

Nuclear plants use nuclear fission reactions to produce electricity. Nuclear fission involves splitting of atoms[276]. An enormous amount of energy is produced during nuclear fission. The speed of the reaction needs to be managed carefully to control the energy. The reactors used in nuclear plants are equipped with systems that can slow down or shut the nuclear reactions when needed.

The heat energy produced from nuclear fission is used to produce steam. The steam drives large turbines to produce electricity using generators.

Uranium is commonly used as nuclear fuel. A certain isotope of Uranium, U-235, is preferred because its atoms can be easily split apart. Nuclear fuel has extremely high energy content. Consequently, nuclear fuel is required in tiny amounts. A typical nuclear power plant only requires about 300 tons of fuel each year[277]. For reference, a coal plant requires about 2.5 million tons of coal to generate the same amount of electricity.

Major advances related to nuclear technology have been in made in recent decades to address safety and cost concerns[278].

Advantages and Challenges

Nuclear power has a midrange-to-high upfront cost and a midrange total lifetime cost compared to other options[279,280,281,282]. The high-capacity factor of nuclear power plants ensures that the costs are lower than expected for a very complex technology[283]. The costs are very low for extending life of an existing nuclear power plant.

Nuclear technology can produce electricity 24X7. Thus, it has a convenience advantage.

Nuclear power plants produce radioactive waste, which requires special handling for transportation and disposal.

Nuclear projects are challenged by a strong negative perception amongst the general population[284]. Although undeserved, the general belief is that the technology is dangerous[285]. This perception is deep-rooted because of the widely publicized nuclear incidents–such as the Fukushima (2011), Chernobyl (1986) and Three Mile Island (1979) accidents[286].

Biomass Power

Biomass is organic matter from plants or animals. Examples of biomass include wood, agricultural crops and manure.

Biomass was the largest source of global energy for several centuries. Even today, many developing countries use large quantities of biomass for heating and cooking. Currently, the use of biomass for electricity production is very small. Biomass power contributes to less than 3% of the total electricity production globally[287,288].

Biomass power has moderate environmental impacts compared to other options in the power sector[289,290,291,292].

Technology Basics

Biomass is burned as a fuel to produce heat energy in a biomass power plant. The heat energy is converted to electrical energy via a turbine-generator combination. Overall, the biomass to electricity process is like the fossil fuels to electricity process.

The biomass process releases comparable amount of CO_2 as fossil fuels. But the CO_2 released from biomass during electricity generation is largely balanced by the CO_2 captured during its growth[293]. Hence, biomass produces lower net greenhouse gas emissions compared to fossil fuels. The greenhouse gas emissions are not completely offset as the life cycle of the biomass process includes other emissions such as those generated during growing, gathering and transportation of biomass[294].

Advantages and Challenges

Biomass power has a midrange upfront cost and a midrange-to-high total lifetime cost compared to other options[295,296,297,298]. The total lifetime cost is sensitive to feedstock costs.

Despite general similarities, costs of biomass power are higher than fossil fuel power because biomass conversion to electricity is more complex. The complexity arises from the challenging properties of biomass feedstock. For example, biomass feedstock has low density and a very high oxygen content.

Like fossil fuels, the electricity from biomass power is available when needed. Thus, it too has a convenience advantage.

Biomass plantations have large water and land requirements. Also, transporting biomass feedstock over long distances is energetically unfavorable because of its low density. Therefore, biomass power plants need to be located close to biomass plantations.

Fossil Fuel Plants with Carbon Capture & Storage (CCS)

CO_2 is a major byproduct from fossil fuel power plants. However, these power plants can produce low-carbon electricity using carbon capture and storage (CCS), i.e., by decreasing CO_2 release into the atmosphere.

CCS technology received very little attention in the past because the financial incentives were lacking. Consequently, it has seen negligible deployment.

Currently, the technology is used to capture less than 50 million tons of CO_2 per year[299]. For reference, coal power plants alone generate around 10,000 million tons of CO_2 per year.

CCS can be used to lower the greenhouse gas emissions from new build natural gas and coal power plants as well as from the operating natural gas and coal plants. Herein, new build natural gas and coal plants with CCS are discussed to facilitate the comparison with other low-carbon technologies.

Technology Basics

Coal and natural gas power plants burn coal and natural gas, respectively, to produce heat energy. A turbine-generator combination is used to convert the heat energy to electrical energy[300]. The product stream from the power plant contains CO_2 and other gases.

CCS for power plant applications involves the following: 1) capture of CO_2 from the gaseous products 2) transportation to the storage location and 3) storage in geological formations.

CO_2 is captured in the power plants via chemical absorption or physical separation. The CO_2 capture systems are typically designed to capture 85 to 90% of the produced CO_2. The capture efficiency depends on the concentration of the CO_2 in the gaseous products and the nature of contaminants. The difficulty of CO_2 capture increases markedly at low concentrations.

The captured CO_2 is compressed prior to its transportation. Compression to a dense phase facilitates its transportation. The storage sites are typically far from the power plants. Pipelines and ships are the favored options for transporting the captured CO_2 over long distances.

The final step involves CO_2 injection deep into underground geological formations to prevent its escape into the atmosphere.

Advantages and Challenges: Natural gas power plants with CCS

The CCS component adds substantial cost to natural gas power plants. For example, the upfront cost for a natural gas power plant with CCS is over twice that of a natural gas power plant.

Natural gas power plant with CCS has a low-to-midrange upfront cost and a midrange-to-high total lifetime cost compared to other options[301,302]. The total lifetime cost is sensitive to natural gas price. Hence, this technology has an advantage in regions with low natural gas price. For example, United States has a midrange total lifetime cost because natural gas is available at a low cost[303].

On a lifecycle basis, CCS reduces greenhouse gas emissions from natural gas power plants by 60 to 80%.

A natural gas power plant with CCS can provide 24X7 electricity. Thus, it has a convenience advantage.

Advantages and Challenges: Coal power plants with CCS

Coal power plants have much higher upfront costs compared to natural gas power plants. Similarly, coal power plants with CCS also have a much higher upfront cost than natural gas with CCS plants. Overall, coal power plant with CCS has a high upfront cost and midrange-to-high total lifetime cost compared to other options[304,305].

The competitiveness of the total lifetime cost depends on the coal price. For example, the United States has a relatively high coal price. Therefore, the technology has a high total lifetime cost in the United States.

CCS reduces greenhouse gas emissions from coal power plants by 60 to 80% on a lifecycle basis. Even so, coal power plants with CCS have higher emissions than other low-carbon technologies (**Figure 6.1**).

The electricity from a coal power plant with CCS is available when needed. Hence, it has a convenience advantage.

Geothermal Power

Geothermal energy is the natural heat of the earth. The slow decay of radioactive particles in the earth's core produces heat[306]. Geothermal power belongs to the renewable category because its source of energy–heat is continuously produced in the earth.

The earth's interior contains an enormous amount of heat energy. But only a small fraction is practically accessible[307].

Geothermal energy was used in North America over ten thousand years ago[308]. But the use of geothermal energy to produce electricity is relatively recent. The first commercial geothermal power plant was operated in Larderello, Italy in 1913[309].

United States currently produces the most electricity–17 billion kWh–from geothermal energy in the world[310]. However, geothermal power represents less than 0.5% of the total electricity generated in the United States.

New Zealand, Iceland, El Salvador and Kenya produce over 15% of their national share of electricity via geothermal power[311]. But geothermal power plants contribute to less than 1% of the global electricity[312].

Geothermal power has low environmental impacts compared to other options in the power sector[313,314].

Technology Basics

Geothermal resources can be used for heating, cooling and electricity production[315]. A geothermal resource requires heat, fluid and permeability to produce electricity[316]. A conventional hydrothermal resource is a high-quality resource because it contains all three elements. Historically, geothermal power plants have used conventional hydrothermal resources because of this major advantage.

A geothermal power plant uses steam from the geothermal resource to generate electricity via a turbine-generator combination. Geothermal electricity generation requires water or steam at high temperatures, 150 to 370° C.

There are three main types of geothermal power technologies: dry steam, flash steam, and binary cycle. Technology selection depends on the state and temperature of the fluid[317]. Currently, flash steam technology is the commonly used technology.

The exclusive use of high-quality resources restricts the deployment scale of geothermal power because of the limited availability of such resources. Enhanced geothermal systems are being considered to produce electricity from the widely available but lower-quality geothermal resources[318].

The lower-quality resources lack the necessary ground water and/or rock permeability. Therefore, enhanced geothermal systems require innovative subsurface engineering and transformation[319]. This increases the complexity and cost of the project.

While conventional geothermal power is an established technology, enhanced geothermal systems are currently in the development and demonstration phase.

Advantages and Challenges

Geothermal power has low costs compared to other options when high-quality resources are available [320,321,322]. However, high-quality resources are scarce. Currently, geothermal power has high upfront and total lifetime costs when considered for wide-scale deployment, i.e., when using lower quality resources[323,324].

Electricity generation from geothermal power plants is not constrained by weather[325]. Electricity from geothermal power is available when needed and thus has a convenience advantage. Also, geothermal power plants have a much smaller land footprint compared to other renewable technologies.

Enhanced geothermal systems–which use low quality resources–are decades away from being considered a mature technology[326]. Therefore, wide-scale deployment of geothermal power is not expected soon. But there is decent potential for geothermal power capacity additions in countries that can use existing high-quality resources[327,328].

...§§§-§-§§§...

7. Low-Carbon Transportation Options

"Transportation is the center of the world. It is the glue of our daily lives. When it goes well, we do not see it. When it goes wrong, it negatively colors our day, makes us feel angry and curtails our possibilities."–***Robin Chase***

The transportation sector contributes to 22% of the greenhouse gas emissions from the global energy sector[329,330,331]. Light duty vehicles, road freight vehicles, aviation, and shipping are the major contributors[332,333].

The largest category in the transportation sector is light duty vehicles. This has led to a focus on decreasing greenhouse gas emissions from light duty vehicles. Examples of light duty vehicles include cars, sport utility vehicles, crossovers, light trucks and minivans.

Battery electric vehicles, biofuels and fuel cell electric vehicles are considered as the low-carbon technology options for the transportation sector[334].

Energy efficiency and energy density are useful characteristics for defining performance.

- Energy efficiency is the percent of energy from the fuel that is available to move the vehicle down the road[335].
- Energy density is the energy available per kilogram of the resource[336].

High energy efficiency and high energy density are desirable characteristics for low-carbon vehicles.

Vehicles can be classified into two categories based on their load and travel range requirements: 1) light duty vehicles and 2) road freight, aviation, and shipping. Light duty vehicles carry much lighter loads and travel shorter distances.

The advantages and challenges are influenced by the vehicle category. The two categories are, therefore, discussed separately.

Battery Electric Vehicles (BEVs)

Conventional vehicles are powered by liquid hydrocarbon fuels. In contrast, BEVs are powered by electricity stored in batteries.

BEV technology was invented in the early nineteenth century[337,338]. The invention of rechargeable batteries and powerful electrical motors provided a boost to the technology in the1870s. BEVs became popular following these inventions.

But conventional vehicle manufacturers also made significant innovations towards the end of the nineteenth century. These innovations resulted in cost and convenience advantages for conventional vehicles over BEVs. Consequently, the interest in BEVs was short-lived.

BEVs have seen renewed interest in recent years because of climate change concerns. Although sales have been increasing rapidly, BEVs represent a tiny fraction of the global vehicles. For reference, BEVs currently make up only 1% of the global car stock[339,340].

Technology Basics

Electrical energy is used to power the motor in BEVs. The electrical energy is stored in a stack of batteries, aka battery pack. Li-ion based batteries are commonly used. The batteries are recharged by plugging into an electric power source.

BEVs are three times more energy efficient than conventional vehicles[341,342]. BEVs have fewer components and moving parts compared to a conventional vehicle. So, battery electric fuels have lower fuel and routine maintenance costs.

But batteries have very low energy densities compared to conventional fuels[343]. A Li-ion battery holds about fifty times less energy than gasoline per kilogram. The higher energy efficiency of electric vehicles offsets some of the impact from its poor energy density. But only to a small extent. Hence, BEVs are significantly heavier than conventional vehicles. Vehicle weight is a critical property for the freight vehicles, aviation, and shipping category.

BEVs have a lower travel range compared to conventional vehicles because of inferior energy density. Travel range is the maximum distance that a vehicle can travel after it is fully fueled or recharged. The battery size determines the travel range. For example, Nissan Leaf with a 40-kWh battery has a travel range of 149 miles, while that with a 60-kWh battery has a travel range of 212 miles[344]. For reference, conventional vehicles have a travel range of 400 miles.

Advantages and Challenges: Light duty vehicles

BEVs require expensive batteries for a travel range comparable to conventional vehicles. Hence, BEVs have a high upfront cost. BEVs with a travel range of 250 miles have 40 to 50% higher upfront cost compared to conventional vehicles[345,346,347]. For example, the retail price for 2022 Kona Electric (SEL) is 48% higher than the 2022 Kona (SEL)[348].

The lower fuel and routine maintenance costs of BEVs have a positive impact on the total lifetime cost. Also, BEVs have large fuel savings in countries where gasoline price is considerably higher than electricity price.

Consequently, BEVs have a low-to-midrange total lifetime cost in countries that have a high gasoline price[349,350,351] and a midrange total lifetime cost in countries that have a low gasoline price[352,353,354].

Conventional vehicles emit particulate matter, CO, and greenhouse gases. BEVs have zero tailpipe emissions[355]. However, BEVs are responsible for the greenhouse gases and other pollutants released from the upstream processes. Production of electricity and manufacturing of vehicles are examples of upstream processes.

The carbon intensity of the region's electrical grid defines the greenhouse gas emissions from BEVs. Carbon intensity is the amount of greenhouse gas emissions produced per unit of electricity generated. Electrical grid includes the entire chain–from electricity generation to its distribution to the consumers.

Currently, BEVs produce 50% lower greenhouse gas emissions compared to conventional vehicles on a lifecycle basis[356,357,358]. This estimate is for mid-sized vehicles and is based on a global average carbon intensity of the electrical grid.

BEVs require battery recharging, which presents a major inconvenience. For most of the global population, recharging is not possible at their homes. This is a problem because the availability of charging stations is low compared to gas stations. Also, most public charging stations require over ten times more time for recharging than refueling at gas stations[359,360]. Currently, BEVs at most public charging stations require over an hour for recharging[361]. In contrast, a conventional vehicle requires a few minutes at a gas station.

Newer generation charging stations can recharge vehicles at a faster speed but are still much slower than conventional fuel pumps[362,363].

From a cost perspective, the electricity price at public charging stations is much higher than the residential price.

The travel range of BEVs is substantially shorter than conventional vehicles. A typical BEV has a travel range of 250 miles, while a conventional vehicle has a travel range of 400 miles.

A shorter travel range coupled with slow recharging speeds results in a convenience challenge. Travel range can be increased by increasing battery size. However, the vehicle cost is sensitive to battery size[364,365]. Higher battery size equals higher upfront cost and higher total lifetime cost.

Advantages and Challenges: Road Freight, Aviation and Shipping

Road freight vehicles carry much larger loads and have longer travel range requirements compared to light duty vehicles[366]. This results in heightened challenges.

A longer travel range requires a larger battery size. Batteries store much less energy per unit kilogram compared to conventional fuel. So, large batteries increase the weight of a freight vehicle.

Freight vehicles have restrictions on their total weight[367]. For example, a semi-truck, a common road freight vehicle, has a maximum total laden weight restriction of 80,000 pounds in the United States[368]. The large heavy batteries required for battery electric freight vehicles decrease the payload carrying ability relative to conventional vehicles.

Currently, BEVs in the road freight category have a maximum range of 500 miles. Tesla is marketing vehicles with a 300-mile and 500-mile travel range in this category[369]. For reference, conventional vehicles typically have a travel range between 1000 and 2000 miles[370]. Thus, BEVs require a large compromise in travel range or payload. The negative economic implications are substantial.

Recharging BEVs is several times slower than conventional fueling which aggravates the challenge. Major advances in technology will be required for the battery electric freight vehicles to provide similar performance and convenience to conventional freight vehicles.

Vehicles in the aviation and shipping categories typically have even larger loads and longer travel range requirements. Hence, battery electric applications are challenged to an even greater extent. Consequently, the interest in battery applications for aviation and shipping is limited.

Fuel Cell Electric Vehicles

Fuel cell electric vehicles are powered by the electricity from fuel cells. Hydrogen, the typical fuel in fuel cells, does not occur naturally. It is predominantly produced from natural gas. But it can also be produced by splitting water molecules and biomass gasification.

Hydrogen is used in applications such as petroleum refining, fertilizer synthesis, and food processing. Currently, its use in the transportation sector is trivial.

The first hydrogen fuel cell vehicle was produced by General Motors in the 1960s[371]. The project was terminated because of cost and other practical issues.

Towards the end of the twentieth century, there was a large emphasis on decreasing vehicular emissions. In response, several auto manufacturers launched efforts on fuel cell electric vehicles. However, these efforts failed to popularize these vehicles.

In recent years, fuel cell electric vehicles have found renewed interest because of the emphasis on reducing greenhouse gas emissions.

Technology Basics

In a hydrogen fuel cell, hydrogen reacts with oxygen electrochemically to produce electricity, water and some heat. The electricity produced in hydrogen fuel cells is used to power the motor in a fuel cell electric vehicle.

Hydrogen is classified based on its production method[372]. Hydrogen produced from coal is termed as brown hydrogen, while that from natural gas or petroleum is grey hydrogen. If hydrogen is produced from either coal, natural gas or petroleum and is coupled with carbon capture and storage, it is called blue hydrogen. Hydrogen produced from splitting (electrolysis) of water is called green hydrogen, if renewable electricity is used to split water. Finally, hydrogen produced using nuclear power is called pink or clean hydrogen.

Fuel cell electric vehicles are over two times more energy efficient than conventional vehicles[373,374]. But they are less efficient than BEVs.

Hydrogen is stored in fuel cell electric vehicles using compressed tanks. Fuel cell vehicles have a longer travel range than a typical BEV. For example, Toyota Mirai, a mid-sized fuel cell electric car, has a travel range of 400 miles[375].

Advantages and Challenges: Light duty vehicles category

Fuel cell electric vehicles are substantially more costly than conventional vehicles because fuel cell technology is expensive. The total lifetime costs are also considerably higher because hydrogen fuel costs are currently very high[376]. When compared to available options, fuel cell electric vehicles have a high upfront cost and a high total lifetime cost[377,378].

Like BEVs, fuel cell electric vehicles also have zero tailpipe emissions. The lifecycle greenhouse gas emissions depend on the hydrogen production method. The greenhouse gas emissions are modestly higher than BEVs based on the conventional hydrogen production method[379].

Fuel cell electric vehicles require similar fueling time as conventional vehicles[380]. This is a significant advantage over BEVs. But hydrogen fueling stations are scarce.

Advantages and Challenges: Road Freight, Aviation and Shipping

Fuel cell electric vehicles can travel longer distances and weigh less compared to BEVs because of the superior energy density of hydrogen compared to batteries. Thus, fuel cell electric vehicles are superior in terms of travel range and payload capacity[381]. These are important advantages in this category.

However, high costs and lack of fueling infrastructure are formidable challenges.

Advanced Biofuels

Advanced biofuels use non-edible biomass feedstocks. Edible feedstocks can also be used to produce fuels. But massive use of edible feedstocks is not practical because it will impact food availability[382]. Also, the reductions in greenhouse gases are lower for edible feedstocks because of the large need for fertilizers[383,384].

Consequently, discussions herein are restricted to advanced biofuels[385]. Feedstock examples for advanced biofuels are wood, grass, crop residues, forestry residue, and algae. Advanced biofuels are also known as second generation biofuels. Currently, advanced biofuels have a minuscule share, less than 0.1%, in the transportation sector[386].

Technology Basics

The conversion of biomass to advanced biofuels involves two steps: a) breaking down the biomass to simpler intermediates and b) upgrading the intermediates to the desired fuel product.

Thermochemical and biochemical conversion are the two main processes used for breaking down the biomass[387]. A thermochemical conversion involves high temperatures, while a biochemical conversion involves biological agents such as enzymes. Examples of advanced biofuels include bioethanol, bio-gasoline, biodiesel, bio-butanol, and bio-methanol.

The technology choice for the conversion and upgrading processes depends on factors such as feedstock characteristics and the desired product.

Advanced biofuels can be used directly in conventional vehicles or via minor modifications. This depends on the properties of the biofuel.

Advantages and Challenges: Light duty vehicles

Advanced biofuels are substantially more costly than conventional fuels[388,389,390]. Biomass feedstocks have a lower energy density and more challenging chemistry relative to crude oil. This increases the difficulty of every step associated with the production of advanced biofuels–from growing and harvesting to the conversion[391].

The cost of producing advanced biofuels from biomass is also high because of the poor economy of scale. For reference, the production scale of an average cellulosic ethanol plant is hundred times smaller than an averaged sized oil refinery[392,393].

Because of these challenges, the capital cost per unit of the product is over five times higher for an advanced biofuels facility compared to an oil refinery[394,395,396]. Also, the total lifetime cost is two to three times higher than conventional fuel[397,398,399].

When compared to available options, advanced biofuels have a high upfront cost and a high total lifetime cost[400,401].

Advanced biofuels can reduce greenhouse emissions by about 90%[402], which is markedly superior compared to the current reductions achieved by electric vehicles. Recall, the extent of greenhouse gas emission reductions from BEVs depends on the carbon intensity of the electrical grid.

Advanced biofuels have comparable properties to conventional fuels. They are easily transportable, pumpable and have moderate-to-high energy densities. Consequently, advanced biofuels have a convenience advantage with respect to fueling and travel range.

Advantages and Challenges: Road Freight, Aviation & Shipping

Advanced biofuels have comparable properties to conventional fuels. Therefore, advanced biofuels do not suffer from the travel range, payload and refueling challenges associated with BEVs. These are critical advantages in this category.

However, high costs are a formidable challenge for advanced biofuels.

...§§§-§-§§§...

8. The Debate Around Energy Options

Misinformation rules when crucial information is suppressed.

Social media and news outlets bombard the global population with reports about low-carbon energy options every day. Most reports use localized information and provide biased perspectives.

While localized information is not technically false, it conveys an incomplete message. Consequently, several crucial aspects are ignored when localized information is used by social media and news outlets. This leads to incorrect perceptions or myths about the low-carbon energy solutions–which in turn delays helpful discussions.

These incorrect perceptions need to be examined before discussing a path forward strategy. Herein, a question-and-answer format is used to discuss the key misunderstandings.

Common Issues for Low-Carbon Energy Options

What are the common flaws related to the discussions about low-carbon technologies?

A robust discussion requires the following: a) comparison of the technologies on an apples-to-apples basis, b) inclusion of all crucial information and c) recognition of outliers. The discussion is flawed if these elements are dealt with incorrectly. Unfortunately, this happens routinely. The media platforms are awash with flawed discussions about low-carbon solutions. Some cases are discussed below.

Subsidies are often included in cost comparisons. This is not appropriate for an apples-to-apples comparison. Subsidies are financial support provided by the government and are often temporary. They are based on the whims of the local or national governments. In other words, subsidies are not integral to the technology. Inclusion of subsidies results in misleading cost comparisons and thereby sub-optimal decisions.

Consider an example where a state provides a 15% subsidy for purchasing land for utility solar power plants, and 30% subsidy for purchasing equipment for residential solar. The inclusion of subsidies will not allow a valid cost comparison between the two technologies.

Effectively, it will disallow assessment of the cost-efficiency of the technologies.

The vehicle cost comparison between different technologies is often reported incorrectly in media. This is because many analyses do not eliminate the external factors that can influence the cost.

Examples of external factors in this case are body style, cargo volume and special features. How to minimize the influence from these external factors? This can be achieved by considering different vehicle technologies having the same brand, model and trim. For example, an accurate cost difference between BEVs and conventional vehicle technologies can be obtained by comparing the retail price of Hyundai Kona Electric (SEL) and Hyundai Kona (SEL) for the same production year[403].

The omission of crucial information also leads to flawed discussions. Media articles hype the positive aspects of the popular low-carbon options but omit crucial negative information. This leads to a loss of credibility and delays useful discussions. One example is reviewed below.

BEVs are widely advertised as zero emission vehicles. This gives the impression that BEVs reduce greenhouse gas emissions by 100%. BEVs indeed have zero tailpipe emissions. However, that statement provides incomplete information about its actual emissions.

Lifecycle greenhouse gas emissions from BEVs include those produced during vehicle manufacturing and electricity production. BEVs reduce lifecycle greenhouse gas emissions by 50% when replacing conventional vehicles in a region that has an electrical grid with average global carbon intensity[404,405].

In practice, the extent of reduction depends on the carbon intensity of the local electrical grid. The reduction is low for an electrical grid with high carbon intensity and is high for a grid with low carbon intensity. Such information is crucial because the carbon intensity of the electrical grids vary widely across regions and countries. Unfortunately, this is rarely highlighted in media.

An incorrect focus on outliers is another common mistake. Exceptions can always be found. However, they are statistically unlikely. For example, some high school dropouts have become very successful. However, this is a statistical abnormality caused by unique conditions. In general, individuals who complete high school are more likely to have a successful career than those who do not. It is misleading to hype the success of high school dropouts without noting that it is unusual. Why? Because it leaves the impression that high school education is not important.

In other words, exceptions or information based on unique circumstances are not widely applicable. Similarly, it is wrong to hype costs or other issues about energy technologies that are caused by unique circumstances.

Media often highlights the aspects that arise from unique conditions. But it does not acknowledge the unique conditions. For example, certain renewable projects have very low costs because of perfect location or large financial incentives. Hyping the cost of such projects without highlighting the unique conditions is misleading.

Summary: Three major flaws are commonly associated with discussions about low-carbon technologies. The first flaw relates to invalid comparisons, i.e., comparisons that lack an apples-to-apples basis. The second flaw relates to the cherry picking of information to promote a viewpoint. The third flaw relates to using data that is not widely applicable to either hype or discredit a technology.

The Debate Around Renewable Power

Are renewable technologies such as solar and wind not viable?

The weather conditions dictate electricity generation from solar and wind. However, describing solar and wind technologies as unviable is misleading.

Electrical grids can reliably handle significant levels of solar and wind[406]. Many countries have a history of reliable operation of electrical grids with a 10% or more share of solar and wind[407].

Recall, solar and wind technologies emit miniscule amounts of greenhouse gases compared to fossil fuel power plants. Therefore, solar and wind are important for addressing climate change.

However, solar and wind suffer from a critical intermittency challenge because of their dependence on weather conditions. This challenge has large cost implications for wide-scale deployment.

Summary: Solar and wind cannot generate 24X7 electricity on a standalone basis because of their strong weather dependence. Fortunately, tools are available to address this challenge. Solar and wind power are important solutions for addressing climate change because of their miniscule greenhouse gases emissions and other advantages. However, their intermittency challenge has large cost implications for wide-scale deployment.

Can the cost of solar and wind be directly compared with fossil fuel and other dispatchable technologies?

Solar and wind technologies provide electricity on an intermittent basis, i.e., they lack the functionality to meet 24X7 electricity demand. On the other hand, fossil fuel and nuclear power plants provide 24X7 electricity. The U.S. EIA lists solar and wind in a separate category because of this important distinction[408]. Solar and wind are listed as resource-constrained technologies, while nuclear and fossil fuel power plants are listed as dispatchable technologies.

The intermittency challenge of solar and wind causes imbalance between electricity production and demand[409,410]. Electricity is only generated during certain hours of the day in case of solar. But electricity demand exists around-the-clock, which causes an imbalance. The imbalance is addressed by lowering or increasing the output from the dispatchable power plants[411,412].

Essentially, the dispatchable power plants are forced to sacrifice their performance to accommodate solar and wind power. This forced inefficient use of the dispatchable power plants causes an overall sub-optimization of the electrical grid[413]. This increases the overall electricity cost[414].

The forced curtailing (i.e., restricting) of electricity production from solar or wind is another instance of sub-optimization[415]. For example, California was forced to curtail 1,500,000,000 kWh of electricity production from solar in 2020 because of the imbalance between production and demand[416].

How much is 1,500,000,000 kWh? The global average consumption of electricity per person in 2020 was about 3300 kWh[417]. Thus, in 2020, California was forced to curtail the amount of electricity that would have met the needs of over 400,000 individuals.

The share of electricity generation from solar in California was only about 15% in 2020[418]. Despite these low levels, California was forced to curtail about 5% of its utility-scale solar electricity production in 2020. These forced curtailments are expected to increase significantly with increasing solar deployment[419,420].

Sub-optimization of the electrical grid has a negative impact on the economics[421]. The extent of sub-optimization increases with increasing deployment of solar and wind power in the grid[422]. Consequently, wide-scale deployment of solar and wind is costly.

The Organization for Economic Cooperation and Development (OECD) recently co-published a comprehensive report on this topic[423].

The report provides an example estimate for the cost penalty resulting from the sub-optimization caused by solar and wind. A 10% deployment of solar and wind was found to add a 5% cost penalty to the total electricity cost. Notably, a 50% deployment of solar and wind was found to penalize the total electricity cost by over 40%.

The penalty results from the extra costs related to the deficiencies of solar and wind[424]. These additional costs arise from the variability of power output, uncertainty in power generation and increase in costs for transmission and distribution associated with solar and wind power[425].

The costs that are routinely reported in media do not include the cost penalties. Thus, the reported costs do not reflect the true costs for solar and wind power. Only partial costs of solar and wind power are widely reported in media. Consequently, the reported costs cannot be directly compared with dispatchable technologies.

Media is flooded with articles advertising that solar and wind power is cost competitive with fossil fuel power. Such cost comparisons are misleading unless the shortcomings and related implications are also highlighted. Unfortunately, valid cost comparisons are not discussed in media. By excluding crucial information, these media reports provide an incorrect perception about the true cost of a low-carbon transition.

Summary: Solar and wind cannot generate 24X7 electricity on a standalone basis. Therefore, their costs cannot be directly compared with technologies that can generate 24X7 electricity. The effect of intermittency must be included in a realistic cost comparison. When the intermittency deficiency is included in the cost estimate, the costs for solar and wind power increase significantly. The cost increase is directly proportional to the extent of deployment. Higher solar and wind power in the electrical grid equals higher cost penalty. Unfortunately, most publications focus only on the partial costs of solar and wind power, i.e., they exclude costs related to the intermittency deficiency of solar and wind power.

Why the delay, when some countries already generate most of their electricity from renewable sources?

Certain unique conditions must be satisfied for a country to generate most of its electricity from renewables. Either the country must generate most of its electricity from specific renewable sources such as geothermal or hydrothermal OR it must be able to export and import large amounts of electricity from neighboring countries.

Solar and wind produce electricity intermittently–which causes a supply and demand imbalance. On the other hand, renewable sources such as hydropower and geothermal can provide round-the-clock electricity. Therefore, the countries that use hydropower or geothermal for most of their electricity requirements will not have a supply and demand problem.

Why is this only possible for a few select countries? Because most countries can generate only a small percent of their total electricity from hydropower and geothermal. Hydropower and geothermal projects require suitable locations for achieving low electricity costs[426,427,428]. Unfortunately, suitable locations–i.e., those with very specific characteristics–are scarce.

Hence, only a few countries have adequate hydropower and geothermal resources to meet their electricity needs. Norway and New Zealand are examples in this category[429]. Norway generates about 99% of its electricity from renewable sources, over 90% of which is from hydrothermal. New Zealand generates about 80% of its electricity from renewable sources, majority of which is from hydrothermal and geothermal.

A country can have a major share of its electricity from solar and wind. But for this, it must be able to export and import large amounts of electricity, as needed, from neighboring countries.

Why is this possible only for a few select countries? Because some very uncommon conditions must be satisfied–as is the case for Denmark.

Denmark generates about 50% of its electricity from wind[430]. The intermittent electricity production from wind causes substantial mismatch in supply and demand. According to a report co-authored by the Danish Energy Agency, the neighboring countries such as Norway and Sweden absorb the excess electricity when wind speeds are high in Denmark[431]. The neighboring countries also provide electricity when wind speeds are low.

The energy systems of neighboring countries, thus, serve as an energy storage system for Denmark's renewable power[432]. In other words, these neighboring countries play a critical role in balancing Denmark's electricity supply and demand.

Overall, the situation works out because a) Denmark has very low electricity consumption[433], b) The energy systems of the neighboring countries have the necessary characteristics to serve as an energy storage system for Denmark's renewable power[434], and c) Denmark is robustly connected with neighboring countries which allows smooth export and import of electricity on an as-needed basis[435].

Denmark would have to curtail electricity production or face blackouts in absence of these unique conditions. For example, if Denmark's electricity consumption had not been very low, the limited electricity available from the neighboring countries would have been unable to meet the demand-supply deficit resulting from low windspeeds.

So how small is Denmark's electricity consumption? It is 6% of Germany's electricity consumption and less than 1% of the United States consumption[436]. In fact, the United States loses five times more electricity during transmission and distribution than Denmark's total electricity consumption[437].

Summary: The countries that currently have a major share of their electricity production from renewable sources have unique circumstances. These circumstances do not apply to most of the global electricity production. Thus, the media hype around such countries is misleading.

Can cost of solar and wind power become lower than fossil fuel power for 24X7 electricity production?

Many reports suggest that solar and wind power are already cheaper than fossil fuel power. The claims are incorrect because the comparisons are invalid.

Valid comparisons require an apples-to-apples technology cost analysis. Fossil fuel technologies provide 24X7 electricity, which is a must-have characteristic. Therefore, 24X7 electricity must be the final product from all the technologies under consideration.

Solar and wind power can also produce 24X7 electricity by including energy storage and/or overbuilding solar and wind power plants and/or extending the transmission infrastructure. The extra costs must be included for a valid comparison with fossil fuel power[438].

Basic scientific principles can inform whether the cost of solar and wind power can realistically fall below fossil fuel power[439]. What are basic scientific principles? Those that do not change with time or new findings.

Energy from the sun or solar energy is the primary energy source for solar power and wind power. Solar energy is also the primary energy source for fossil fuels. Thus, solar, wind and fossil fuel technologies have an identical source of primary energy.

Also, these technologies have an identical final point or final product, which is 24X7 electricity. Thus, solar, wind and fossil fuel power have an identical primary energy source and final point.

Which technology will have the lowest cost? The one that requires the least human activity and resources. For example, moving cargo from Houston to Mumbai requires more human activity and resources than moving the same cargo from London to Mumbai. Consequently, the costs are higher for moving cargo from Houston to Mumbai.

For the power sector, examples of human activity include mining, transportation of materials, pretreatment of materials, land preparation, and manufacturing. Examples of resources include land, water, minerals, energy and materials.

What decides the requirement for human activity and resources?

The amount of support from nature decides the required human activity and resources. A technology with a small helping hand from nature will require higher human activity and resources compared to a technology that has a large helping hand. Thus, the technology that receives the most helping hand from nature will require the least human activity and resources and have the lowest cost.

How do the technologies compare in terms of support from nature?

First, we will consider fossil fuel power technologies.

The solar energy captured by ancient plants and organisms has been converted to fossil fuels (**Figure 8.1**). Specifically, nature has transformed the dilute and intermittent solar energy into high density fossil fuels that are available 24X7 for energy production. Nature has enabled this transformation process by applying heat and pressure on the ancient plants and organisms in the earth's crust over millions of years[440]. Because of this massive helping hand from nature, fossil fuels can be rapidly extracted, easily transported, stored and converted to usable energy. Thus, nature has equipped fossil fuels with the ability to efficiently provide 24X7 electricity. Essentially, nature has ensured that fossil fuel technologies have a low requirement for human activity and resources for producing 24X7 electricity.

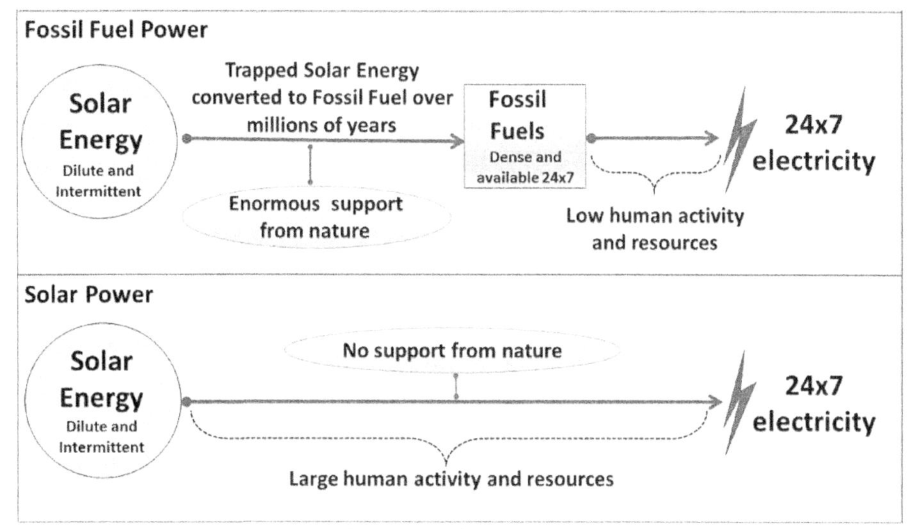

Figure 8.1: Pathways for fossil fuel power and solar power.

We will consider solar power next. Unlike fossil fuels, there is no helping hand from nature for solar power because solar energy is its direct energy source for generating electricity. Correspondingly, solar power requires high human activity and resources for producing 24X7 electricity.

Solar energy has two major challenges. Specifically, solar energy is an intermittent and extremely dilute source of energy[441,442]. These challenges are not lessened in case of solar power because there is no helping hand from nature.

Intermittency is a major challenge for 24X7 electricity production. Therefore, intermittency has large cost implications.

Characteristics such as energy density and power density provide information about how dilute the energy source is[443,444,445]. Based on any reasonable metric, solar and wind are more than thousand times dilute compared to fossil fuels.

The dilute nature of solar energy is a significant challenge. Why? Because it is difficult to harness energy from an extremely dilute energy source[446]. For example, consider the challenges related to fishing in a lake that only has a few fish at any given time.

Next, we will consider wind power. Wind energy is the energy source for wind power.

Recall, nature acts upon solar energy and transforms it to wind energy. However, the helping hand is relatively small. How do we know that? Because wind energy remains an intermittent and a dilute source of energy[447,448].

The required human activity and resources for wind energy are considerably larger than fossil fuels. Why? Because, in case of fossil fuels, nature has eliminated the intermittency and diluteness challenges. Correspondingly, fossil fuels need substantially lower human activity and resources compared to wind for providing 24X7 electricity. This, in turn, results in significantly lower technology costs for fossil fuels.

Thus, basic scientific principles dictate that the cost of solar and wind power will remain higher than the cost of fossil fuel power for producing 24X7 electricity[449].

This discussion can be extended to other renewable power technologies that have solar energy as the primary source. For example, biomass power also has the same primary energy source and final product as solar, wind and fossil fuel power. Biomass power does not have as much support from nature as fossil fuels. Recall, fossil fuels were formed from biomass and other organisms after millions of years of heat and pressure treatments. Consequently, biomass power requires higher human activity and resources than fossil fuels and thereby has higher costs[450,451,452,453].

Currently, the costs of solar and wind with energy storage are very high compared to fossil fuels for producing 24X7 electricity. For example, the upfront cost of solar with battery energy storage is currently over five times higher than natural gas power[454,455].

Solar, wind and energy storage costs will decline with technology advances and increasing deployment–but only up to a point. Despite the decline, the technology costs of solar and wind power will remain significantly higher than fossil fuel power because of the higher requirements of human activity and resources.

Summary: Solar, wind and fossil fuel power technologies have an identical primary energy source (solar energy) and final product (24X7 electricity). Consequently, the amount of support from nature defines the human activity and resources required to produce 24X7 electricity from these technologies. A technology with a smaller helping hand from nature will require higher human activity and resources and vice versa. Because of the drastically smaller support from nature, solar and wind power technologies require substantially higher human activity and resources compared to fossil fuel power technologies. Higher human activity and resources equal higher cost. Consequently, solar power and wind power will have higher costs than fossil fuel power for producing 24X7 electricity in the foreseeable future.

Don't the learning curves indicate that solar and wind costs will soon become lower than fossil fuel power[456]?

Learning curves are not derived from basic scientific principles. They are mathematical curves that show how technology costs have changed with time. Costs of solar, wind and batteries have declined rapidly over the past couple of decades. Therefore, learning curves for these technologies look steep.

Steep learning curves are expected during the period in which the deployment of a technology proceeds from very low to moderate levels. Why? Because large improvements are expected during this period from the rapid technology advances and exploitation of economy-of-scale.

As the deployment increases and reaches a certain threshold level, the technology becomes mature. It transforms from a suboptimal state to a near-optimal state. Consequently, there is smaller room for improvements after the technology reaching maturity.

Correspondingly, cost improvements are small when the technology becomes mature. The technology costs at maturity are mainly determined by the human activity and resource requirements[457].

Fossil fuel technologies reached maturity several decades ago, i.e., they have been at a near-optimal state for several decades. Hence, cost improvements have been small.

In contrast, cost improvements have been large for technologies such as a solar, wind, and batteries because they have yet to reach maturity[458]. The cost improvements will occur rapidly only until the technologies approach a mature state.

The amount of human activity and resources requirements will determine the cost at maturity for these technologies. The human activity and resource requirements are substantially lower for fossil fuel power because of the large support from nature. Consequently, the cost of renewable power will not become lower than fossil fuel power[459,460].

Summary: Fossil fuel technologies have been close to their optimal state for several decades. Hence, they have had small cost improvements. Solar and wind power have had large cost improvements because these technologies were far from their optimal state. The cost improvements will slow down once these technologies reach closer to their optimal state. The human activity and resource requirements determine the costs for technologies at their optimal state. Based on basic scientific principles, fossil fuels require substantially lower human activity and resource requirements compared to solar and wind power. Consequently, the cost of solar and wind power for producing 24X7 electricity will not become lower than fossil fuel power.

Why is there so much debate about green hydrogen?

Green hydrogen has received significant attention in recent years. While some believe it is an excellent solution, others think very poorly of it. Why? Because it has important advantages and difficult challenges.

First, we will consider the advantages.

Green hydrogen can address the intermittency shortcoming of solar and wind power. Existing energy storage technologies have limitations in terms of storage duration or site availability[461,462]. Green hydrogen does not suffer from these challenges. This makes it an interesting option for long duration storage.

Green hydrogen can be used to decarbonize difficult-to-electrify sectors[463]. For example, hydrogen vehicle technology has certain important advantages for road freight vehicles compared to BEV technology. Also, green hydrogen can potentially replace fossil fuels for the difficult-to-electrify heating applications in the industrial sectors.

What about the challenges?

A key challenge is its lower round-trip efficiency compared to other energy storage technologies. Round-trip efficiency is the percent of energy stored that can be later retrieved.

The use of green hydrogen for energy storage involves two steps. In the first step, the low-carbon sources such as solar or wind are used to generate hydrogen via electrolysis of water. In the second step, the green hydrogen is used as a fuel to generate electricity.

There is significant energy loss in both the steps. The combined energy losses result in a round-trip efficiency of 35% for green hydrogen[464]. For reference, the round-trip efficiency of pumped storage hydropower is 80%[465,466]. The round-trip efficiency for Li-ion battery technology is 85% and that of flow battery technology is between 60 and 80%[467]. Thus, green hydrogen has a much lower round-trip efficiency compared to other energy storage options. But unlike most energy storage options, green hydrogen does not suffer from a large cost penalty for long durations of storage[468,469].

The electrolysis and other associated equipment have high upfront costs. The cost disadvantage for electrolysis becomes evident when it is compared with conventional production cost of hydrogen. The total lifetime cost of producing hydrogen by electrolysis is currently three to seven times higher than by the conventional method[470].

Transporting hydrogen is also a challenge because of it chemical and physical properties. This adds to the total lifetime cost. Also, the low round-trip efficiency increases the total lifetime costs. Overall, green hydrogen has high to very high total lifetime costs compared to other options.

Efforts are underway to reduce costs. The viability of green hydrogen will depend on how much and how soon the cost reductions can be achieved.

Summary: Green hydrogen can serve as a versatile energy storage component for solar and wind power. More importantly, it can decarbonize the difficult-to-electrify sectors. Thus, green hydrogen offers some critical advantages. However, it also suffers from major challenges such as high costs and low-round trip efficiency for energy storage. Supporters of the technology find it very promising because of the important advantages. On the other hand, opponents believe that the technology has no potential because of the major challenges.

How much subsidies have renewables received?

Renewables have been receiving substantial subsidies since several years[471,472,473,474]. Most subsidies have been centered around the popular renewables–i.e., solar power, wind power, biomass power and biofuels.

The primary goal of the subsidies is to encourage the deployment of renewables. Example types of subsidies and programs supported by global governments include:

- Tax benefits.
- Financial assistance awards made directly to the recipients.
- Financial assistance awards for R&D.
- Financial support for technologies that are unable to obtain private financing because of high risks.
- Programs such as renewable mandates, net-metering, and feed-in-tariffs[475].

A recent report by the International Renewable Energy Agency (IRENA) gives a feel for the magnitude of global subsidies received by renewables[476]. Solar power, wind power, biomass power and biofuels received global subsidies worth 166 billion dollars in 2017. Note, these popular renewables contributed to less than 5% of the total energy consumed that year[477].

Thus, the popular renewables have received large subsidies. These subsidies have played a significant role in their growth.

Summary: Popular renewables such as solar, wind, and biofuels have been receiving substantial subsidies since several years. For example, the popular renewables received 166 billion dollars in 2017 even though they represented less than 5% of the total energy consumed. These subsidies have a played an important role in their rapid growth.

The Debate Around Fossil Fuel Technologies

Is it appropriate to lump natural gas power plants with coal power plants?

Natural gas technology is considerably superior to coal technology[478,479]. Natural gas power plants generate fewer greenhouse gases and air pollutants compared to coal power plants. Natural gas technology has the lowest upfront cost in the power sector. Also, the total lifetime cost of electricity is low for countries with cheap natural gas.

Coal currently contributes to roughly 35% of the global electricity production[480]. Coal used in power applications produce 10 billion tons of CO_2 per year–which accounts for over 70% of the total greenhouse gas emissions from the power sector.

When an existing coal power plant is replaced by a natural gas power plant, the greenhouse gas emissions are reduced by 50%[481,482]. For reference, BEVs reduce lifecycle greenhouse gas emissions by 50% at a location that has an electrical grid with global average carbon intensity.

As CO_2 capture and storage technology (CCS) matures, the natural gas power plants can be retrofitted with CCS to further reduce greenhouse gases. This can reduce greenhouse gases by an additional 70% reduction[483].

Natural gas technology has a cost and convenience advantage over many low-carbon technologies. The convenience advantage arises from its ability to provide 24X7 electricity.

Thus, the replacement of coal power plants with natural gas power plants is an intermediate solution for decreasing greenhouse gas emissions[484,485].

Summary: Natural gas power plants are superior to coal power plants in terms of upfront cost, pollution and greenhouse gas emissions. Natural gas technology has a cost and convenience advantage over many low-carbon technologies. Moreover, the natural gas power plants can be retrofitted with CO_2 capture and storage technology to further reduce greenhouse gases. Thus, replacement of coal power plants with natural gas power plants is a notable intermediate solution for decreasing greenhouse gases. The lumping of these technologies disallows robust path forward discussions.

Do fugitive emissions cancel the advantages of natural gas power?

Some methane is lost because of leakage during the production and transportation of natural gas. Recent studies have proposed that these fugitive methane emissions could be substantially higher than previously estimated. Based on these studies, some reports have suggested that replacement of existing coal power plants with natural gas power plants will result in little-to-no reduction in greenhouse gas emissions. Is this true?

Two independent credible sources–United States National Renewable Energy Laboratory (NREL) and United Nations Economic Commissions for Europe (UNECE)–have provided recent updates about lifecycle greenhouse gas emissions from different power plant technologies[486,487].

The NREL report has considered around three thousand published lifecycle studies related to electricity generation. The UNECE report includes an updated discussion on fugitive methane emissions.

These reports confirm that the replacement of existing coal power plants with natural gas power plants will decrease lifecycle greenhouse gas emissions by 50%. Thus, replacing existing coal plants with natural gas power plants can substantially reduce greenhouse gases.

Furthermore, regulations can cost-efficiently decrease fugitive emissions. In fact, the International Energy Agency estimates that over 40% of the reductions could be achieved at no net cost[488]. The value of the captured gas would compensate for the abatement cost.

Fugitive emission reductions would further decrease lifecycle greenhouse gas emissions from natural gas power. Hence, a reduction in fugitive emissions should be targeted with high priority.

Summary: Recent studies from two credible sources–NREL and UNECE–have provided updates about lifecycle greenhouse gas emissions from the power sector. These studies include the impact of fugitive emissions. Both studies confirm that replacement of existing coal power plants with natural gas power plants will decrease lifecycle greenhouse gas emissions by 50%. The fugitive emissions can be cost-effectively reduced via regulations. Such emission reductions should be targeted to boost the advantages of natural gas technology.

What does science say about the impact of fracking on water quality?

Fracking, aka hydraulic fracturing, is used to release oil and natural gas from rocks that are impermeable. In this process, fluids are pumped at high pressures to fracture the rocks[489].

Water contamination has been a specific concern about the fracking technology [490].

How substantial is the impact on water quality in practice? To answer this, we will discuss the conclusion from the most reliable sources available.

The U.S. Environmental Protection Agency (EPA) is open-ended in their 2016 report about the impact of fracking on water supply[491]. The report concludes that fracking can impact the water quality under certain circumstances. It lists the various circumstances. However, the report does not provide any concrete conclusions about frequency and severity of impacts. Interestingly, a draft copy of the report issued in 2015 had specifically concluded that the impact was low[492].

The U.S. Geological Survey (USGS) is the largest water, earth, biological science and civilian mapping agency in the United States[493]. Specifically, the agency collects, monitors, analyzes and provides scientific information about natural resources, issues, conditions and problems.

The United States has vast experience in producing oil and natural gas from fracking within its shores[494]. Consequently, the USGS is uniquely positioned to study environmental issues related to fracking.

Over the years, the USGS has dedicated large resources to studying the environmental impact from fracking[495,496]. Their studies show that fracking has not caused widespread contamination of water supplies[497,498,499].

Essentially, the USGS studies show that fracking does not cause excessive problems, when operations are conducted properly. Expectedly, incorrect operations can cause problems[500]. This is a challenge for most industrial operations and is not limited to fracking.

Summary: The U.S. Geological Survey (USGS) is uniquely qualified to study environmental issues related to fracking because it has the best possible access to data and expertise. Focused studies by USGS show that fracking has not caused any excessive impact on the water quality.

What impact do fossil fuels have on air pollution?

Air pollution has several causes. Some examples are burning of fossil fuels, cooking and heating using biomass, chemicals production, burning of agriculture waste, wildfires and mineral dust.

The World Health Organization (WHO) and the Health Effects Institute (HEI) estimate that air pollution is responsible for seven million worldwide deaths per year[501,502,503]. The estimate is for indoor and outdoor pollution combined.

Literature provides a wide range of numerical values for the deaths from air pollution. Consequently, only self-consistent information from reliable sources is suitable for discussions[504].

According to WHO and HEI, outdoor air pollution annually causes about four million deaths[505,506]. But these studies do not provide a breakdown of the contribution from fossil fuels.

Fortunately, a recent collaborative study between multiple organizations provides the required information[507]. This study estimates the global deaths from outdoor air pollution to be four million every year–which is consistent with the WHO and HEI studies[508].

The study also provides information about the contribution from fossil fuels. Overall, the study finds fossil fuels to be responsible for one million of the total deaths from air pollution. Amongst the fossil fuels, coal is the biggest culprit, contributing to more than half of the deaths, i.e., over half a million deaths per year. Combined, liquid petroleum and natural gas contribute to half a million deaths per year.

Each death is one too much. However, context is critical for a practical understanding. The global deaths from other causes are provided below to provide context.

- Deaths related to smoking tobacco (WHO data): over eight million per year[509]
- Deaths related to obesity (WHO data): about three million per year[510]
- Deaths related to use of solid biofuels for residential energy (WHO, HEI data): over two million per year[511,512]
- Deaths due to road traffic accidents (CDC data): over one million per year[513]

Summary: Studies vary widely about the number of deaths caused by air pollution. Therefore, it is essential to consider self-consistent studies that provide systematic information. Such studies indicate that air pollution from coal causes over half a million deaths annually, while the air pollution from natural gas and liquid petroleum combined causes half a million annual deaths. For reference, smoking tobacco causes over eight million deaths and use of solid biofuels for residential energy causes over two million deaths annually.

Why have fossil fuel energy costs fluctuated widely over the past decades?

Energy cost has two components: technology costs and external costs. Basic scientific principles define the technology costs[514]. The actions of the society or natural disasters determine external costs.

Supply disruptions have mainly caused the energy cost fluctuations observed over the past decades. Historical examples of society-caused supply disruptions include wars and inadequate investments for energy production[515]. A recent example of war is the invasion of Ukraine by Russia. This war caused a large increase in the global costs of fossil fuels.

Summary: The historical fluctuations in fossil fuel costs have been primarily caused by disruptions in energy supply and demand. Actions of the society or natural disasters have caused supply disruptions. Examples of societal actions include wars and inadequate energy investments.

Do fossil fuels receive several trillion dollars in subsidies each year?

Subsidies are financial incentives from the government. According to the United Nations Development Programme (UNDP), fossil fuels receive about 425 billion dollars per year in global subsidies[516,517]. For reference, 1000 billion dollars = 1 trillion dollars.

The goals of these subsidies–as stated by global governing bodies–has been energy security and relief from energy price for low-income households. Thus, global governments consider these as energy subsidies and not fossil fuel subsidies. Fossil fuels have contributed to over 80% of the global energy demand[518]. Correspondingly, fossil fuels have received most of the subsidies[519].

Where does the several trillion-dollar subsidies claim come from? According to a series of reports from the International Monetary Fund (IMF), the costs associated with external impacts from fossil fuels are also subsidies[520,521]. For example, costs related to air pollution and climate impacts are considered as subsidies in the report. The IMF report estimates the cost from the external impacts to be several trillion dollars per year. That estimate is the origin of the claim.

The claim has significant credibility problems as discussed below[522].

Historically, the costs from external impacts have been mitigated via regulations and other policies. The cost-efficiency for such efforts is very high. For example, U.S. EPA estimates that the cost of air pollution control technology is 30 times lower than the health costs arising from air pollution[523]. Thus, a 35-billion-dollar investment in air pollution control technology can eliminate a trillion-dollar health cost. In other words, the external cost figures in the report are highly exaggerated.

The report includes road congestion, road accidents and road damage as an external impact from fossil fuels[524,525]. That is like saying the agriculture industry is responsible for the growing obesity in the global population. According to the report, road congestion, accidents and damage represent the costliest external impact from gasoline.

The costs discussed in the report are exaggerated and misleading. Most media reports do not discuss these critical facts. They only advertise that fossil fuels receive several trillion dollars of subsidies.

Summary: Governments have historically subsidized energy costs with the stated goal of providing relief from energy price for low-income families. Fossil fuels have received most of these subsidies–425 billion dollars per year–because fossil fuels have dominated global energy. Some reports claim that fossil fuels receive several trillion dollars in subsidies annually. These claims are based on the supposed external costs related to fossil fuels. However, these claims lack credibility. For example, the largest subsidy for gasoline is claimed to be the external costs associated with road congestion, traffic accidents and road damage. That is like claiming the agriculture industry is responsible for the external costs arising from the obesity epidemic.

The Debate Around Transportation Solutions

Apart from electric vehicles, what options can reduce greenhouse gases from light duty vehicles?

Hybrid electric vehicles and expansion of mass transit are two established solutions for reducing greenhouse gases from light duty vehicles.

Hybrid electric vehicle technology is based on a minor modification to conventional vehicle technology. In case of hybrid vehicle technology, an electric motor is included along with an internal combustion engine. The electric motor uses energy stored in a battery, which is charged by the internal combustion engine and through regenerative braking[526].

Hybrid electric vehicles have 10% higher upfront costs and lower lifetime costs compared to conventional vehicles[527,528,529]. For reference, the retail price of 2022 Toyota Camry hybrid (LE) is 8% higher than a 2022 Toyota Camry (LE)[530]. Overall, hybrid electric vehicle technology has low-to-midrange upfront costs and low total lifetime costs compared to other options in the light duty transportation sector.

Hybrid electric vehicles reduce lifecycle greenhouse gas emissions by 20 to 30% when replacing conventional vehicles[531,532,533]. The reduction is because of the superior energy efficiency of hybrid electric vehicles[534].

Recall, BEVs currently reduce lifecycle greenhouse gas emissions by 50% when they replace conventional vehicles[535]. The actual reduction depends on the carbon intensity of the local electrical grid. It can range from a small reduction for a grid with high carbon intensity to a high reduction for a grid with low carbon intensity[536].

Like conventional vehicles, hybrid vehicles use gasoline or diesel and do not require external charging. Hence, hybrid electric vehicles can be conveniently fueled and have long travel ranges. Overall, hybrid electric vehicles have a cost and convenience advantage over BEVs and advanced biofuels.

Mass transit, aka public transportation, involves the sharing of a vehicle by many passengers. Examples include buses, shuttle vans and rail. The vehicle occupancy of the mass transit vehicle defines the greenhouse gas emission reductions. Vehicle occupancy is the percentage ratio of the actual number of passengers to the maximum passenger capacity. Higher vehicle occupancy equals greater greenhouse gas reductions.

A study by the U.S. Department of Transportation showed that a reasonable vehicle occupancy can decrease greenhouse gases sizably[537]. For example, mass transit vehicles at 70% occupancy can reduce the greenhouse gases by at-least 60% compared to personal (conventional light duty) vehicles[538]. Thus, expansion of mass transit can substantially decrease greenhouse gas emissions from the light duty vehicles sector.

Mass transit has a low cost because of the monetary savings from the drastically reduced total vehicle miles[539,540]. The total vehicle miles are low for mass transit because many passengers travel in the same vehicle.

In contrast, transporting an identical number of passengers requires many light duty vehicles in case of personal transportation. Hence, personal transportation results in much higher total vehicle miles.

Mass transit has low upfront and low total lifetime costs because of low fuel and maintenance costs and fewer vehicles[541]. Mass transit also reduces air pollution, road congestion, traffic accidents and energy consumption because of the drastically lower total vehicle miles. Overall, it has many critical advantages over the use of personal light duty vehicles.

However, mass transit offers lower convenience compared to using personal vehicles. Fortunately, robust policies can help with that.

The expanded use of shuttle vans for mass transit is particularly interesting from a practical viewpoint. For example, shuttle vans do not require new infrastructure like rail. They have a low upfront cost and an optimal passenger capacity for maintaining high occupancy levels[542,543]. Therefore, shuttle vans can be used on multiple routes with high frequency and yet can maintain high vehicle occupancy.

Summary: Hybrid electric vehicles and expansion of mass transit are two established solutions for reducing greenhouse gas emissions from light duty vehicles. These solutions have several advantages over light duty BEVs. Expansion of mass transit is a particularly effective solution because of its many critical advantages compared to personal transportation.

Are electric vehicles currently a very effective solution for addressing climate change?

Light duty BEVs are hyped as one of the most effective solutions for addressing climate change[544]. What is a very effective solution? One that can decrease greenhouse gases at a lower cost compared to other solutions, while ensuring minimal inconvenience.

The cost and convenience characteristics of BEVs relative to other solutions are discussed below. United States is used as a case study.

Most BEVs sold in the United States are small to mid-sized vehicles, for e.g., cars and small sport utility vehicles[545]. Therefore, the case study focuses on the small to mid-sized vehicle categories.

Lifecycle greenhouse gas emissions from different vehicle technologies for the same model and trims are provided for an apples-to-apples comparison[546,547,548].

- Hybrid vs. Conventional (car): Toyota Corolla Hybrid has 30% lower lifecycle greenhouse gas emissions than Toyota Corolla.

- Hybrid vs. Conventional (sport utility vehicle): Toyota RAV4 Hybrid has 26% lower lifecycle greenhouse gas emissions than Toyota RAV4.
- Battery Electric vs. Conventional (sport utility vehicle): Hyundai Kona Electric has 50% lower lifecycle greenhouse gas emissions than Hyundai Kona.
- Battery Electric vs. Hybrid (station wagon): Kia Niro Electric has 25% lower lifecycle greenhouse gas emissions than Kia Niro Hybrid.

The direct comparison of vehicle technologies shows only a moderate advantage for BEVs over hybrid electric vehicles in the small to mid-sized vehicle categories. This is in good agreement with other recent studies focused on mid-sized vehicles[549,550,551,552].

However, BEVs have high costs for reducing greenhouse gases. BEVs have a 40 to 50% higher retail price than conventional vehicles, while hybrid electric vehicles have only a 10% higher retail price than conventional vehicles. Consequently, BEVs have a threefold higher upfront cost for yearly greenhouse gas reduction compared to hybrid electric vehicles. Evidently, the retail price of BEV technology is currently too high for being considered a very efficient solution.

The BEV technology also has a total lifetime cost disadvantage in the United States[553,554,555]. On an average, the total lifetime cost for BEV technology is about 10% higher than hybrid electric vehicle technology. Hybrid electric vehicle technology has lower total lifetime costs compared to conventional vehicle technology.

Recall, BEVs suffer from a major convenience challenge. For example, BEVs require much longer waiting time at public charging stations compared to fueling conventional vehicles at gas stations[556,557]. Also, BEVs have a 40% shorter travel range compared to conventional vehicles. Hybrid electric vehicles do not suffer from these convenience challenges.

Mass transit is a particularly cost-effective option for decreasing greenhouse gas emissions from the light duty transportation sector. Recall, mass transit can reduce the greenhouse gases from the sector by over 60% in the United States[558]. The total lifetime cost is low for mass transit options such as buses, shuttle vans and rail[559].

Why is there so much misunderstanding about BEVs? Because most costs reported in the media are based on studies that are not equipped to robustly compare vehicle technologies. For example, many studies include subsidies[560]. Recall, subsidies are an external factor. Therefore, inclusion of subsidies does not allow an apples-to-apples comparison between technologies.

The importance of a robust comparison is revealed by three recent studies: the 2019 Insights into Future Mobility study by MIT, an independent 2020 analysis and the 2021 collaborative study involving multiple U.S. National Laboratories[561,562,563]. These studies show a significant cost advantage for hybrid electric vehicles over BEVs in the United States. Also, according to a recent report by McKinsey and Company, BEVs are years away from achieving total lifetime cost parity with conventional vehicles in the United States[564].

The other reason for the misunderstanding about BEVs is the emphasis placed on zero greenhouse gas emissions from the tail pipe. For example, U.S. EPA lists BEVs as zero emission vehicles[565]. This emphasis leads to an incorrect perception that BEVs can currently reduce greenhouse gas emissions by 100%.

Are BEVs more efficient in other countries?

BEVs are competitive in terms of total lifetime costs for greenhouse gas reductions in certain countries. These are countries with high conventional fuel prices and an electrical grid with very low carbon intensity[566,567,568]. Examples include Canada and several European Union countries. In such countries, BEV technology can be a cost-effective solution.

BEVs are not efficient for climate change mitigation in countries that use substantial amounts of coal for electricity production. Recall, BEVs on a global average basis can currently reduce lifecycle greenhouse gas emissions by only about 50%. The actual reduction range depends on the carbon intensity of the local electrical grid.

Countries such as India, China and many developing economies use large amounts of coal for electricity production[569], i.e., the carbon intensity of their electrical grid is high. BEVs provide little-to-no advantage compared to hybrid electric vehicles in terms of greenhouse gas emission reductions in such countries. For example, BEVs decrease lifecycle greenhouse gas emissions by about 25% when replacing conventional light duty vehicles in India, and by about 40% in China[570,571].

Hybrid electric vehicles do not suffer from a convenience challenge. Battery vehicles suffer from a major convenience challenge–which is an obstacle for wide-scale deployment. The convenience challenge of BEVs is an even larger obstacle for developing countries. This is related to the massive costs and other challenges associated with building the necessary infrastructure.

Expansion of mass transit, specifically shuttle vans, is the most effective climate change mitigation solution for most countries. It has low costs, low infrastructure requirements and can substantially decrease road congestion and total energy consumption.

Summary: A very effective solution can decrease greenhouse gases at lower costs compared to other options, while ensuring minimal inconvenience. Battery electric light duty vehicles cannot satisfy either of the conditions. For example, BEV technology has a roughly three times higher upfront cost for yearly greenhouse gas reductions compared to hybrid electric vehicle technology. BEVs suffer from major convenience challenge because of a shorter driving range and very slow recharging times. For example, the charging time for BEVs at most public charging stations is at-least ten times slower than fueling conventional or hybrid electric vehicles at gas stations. The slow recharging time translates to long wait times at public charging stations. Expansion of mass transit has many critical advantages over battery electric light duty vehicles such as lower costs, comparable or higher greenhouse gas emission reductions, lower energy consumption and reduced road congestion.

Are we close to resolving the convenience challenge of BEVs?

The hope is that the inconvenience associated with BEVs will be resolved quickly. As detailed below, such hope is not justified.

Recall, a technology is said to have a convenience challenge, if it has a substantial disadvantage relative to the incumbent technology. BEVs have two major disadvantages: very slow charging time and short travel range.

First, we will consider the charging issues.

Residential or at-home charging is convenient. However, a major fraction of the general population must rely on public charging stations because it lacks convenient access to residential charging. According to a study by National Renewable Laboratory, over 60% of the U.S. population currently lacks access to convenient residential charging[572].

Detached single-family homes are particularly suited for convenient residential charging. Despite the United States having a high proportion of detached single-family homes, majority of its population lacks access to convenient residential charging. The situation is much worse in other countries because of the smaller proportion of detached single-family homes.

Access to efficient public charging is also important for people with convenient residential access. For example, for long distance travel or the instances when the vehicle is not charged at the residence.

The importance of public charging stations is noted in several studies. According to a study by the International Council on Clean Transportation, the United States will require 2.4 million public or work-place chargers for 26 million electric vehicles[573]. For reference, the United States currently has less than 0.25 million public or work-place chargers[574,575]. Consequently, a phenomenal increase in the number of chargers will be required if the 250 million conventional light duty vehicles in the United States were to transition to electric vehicles[576].

What is the current situation with public charging stations?

Public chargers are of two types: Level 2 and Level 3. These chargers are primarily distinguished by their charging speeds[577]. Level 2 chargers require few to several hours for providing a 200 miles driving range. Level 3 chargers, aka DC Fast Chargers, require between 15 to 60 minutes for providing a 200 miles driving range. For reference, fueling a conventional or hybrid vehicle requires less than five minutes for a 400 miles driving range.

Level 2 chargers are inconvenient because of extremely long wait-times. Even the wait-time required for Level 3 chargers is several times more than gas stations. Although the inconvenience is mitigated to a certain extent for Level 3 chargers, it is still significant.

Most public chargers in the United States and Europe are Level 2 chargers. According to the U.S. Department of Energy, 80% of the public chargers in the U.S. are Level 2[578]. Also, 89% of the public chargers in Europe are Level 2 according to European Automobile Manufacturers' Association[579].

Level 2 chargers dominate the public charging landscape because Level 3 chargers have very high upfront costs. For example, Level 3 chargers can have over ten times higher upfront costs compared to Level 2 chargers[580].

While Level 3 chargers can reduce the wait-time, they are expensive. Electricity costs at Level 3 public charging stations can be over twice that of residential charging[581,582,583,584,585]. This leads to another problem.

Total lifetime cost estimations for BEVs are typically based on the low residential electricity price, for example, $0.13/kWh in the United States[586,587,588]. Such estimations include a large fuel cost advantage for BEVs. When electricity prices are considered from Level 3 chargers, the fuel cost advantage for BEVs decrease substantially. In fact, a recent study by the Anderson Economic Group shows that fueling costs for BEVs can be higher than conventional vehicles when all costs are considered[589].

Electricity costs play an important role in defining total lifetime costs. If electricity price from public charging stations is included–as it should be–the total lifetime cost for BEVs increases substantially. Thus, the improved convenience of Level 3 chargers comes with an increase in total lifetime costs. This increases the cost challenge for BEVs.

Consequently, two major advances in charging technology are required. Electricity price from Level 3 chargers must decrease significantly. Simultaneously, the charging time must decrease to less than five minutes for providing a 200-mile driving range.

A big risk to wide-scale deployment of BEVs is to be locked into the super inconvenient Level 2 public chargers or the expensive Level 3 chargers. Only early adopters are likely to accept substantial inconvenience and high costs. Early adopters are the small percent, less than 15%, of the population, that is anxious to try new technologies.

Thus, many consumers will likely not adopt BEVs. This risk can be avoided by allowing time for advancing the charging technology prior to deploying public chargers extensively.

What about travel range issues?

Travel range is important because it dictates how long the vehicle can travel before refueling. The average travel range for BEVs is 250 miles, while the average travel range for conventional vehicles is 400 miles[590].

To achieve the same convenience levels as conventional vehicles, the BEVs must achieve a 400-miles travel range. Higher travel range can be obtained for BEVs but at a substantially increased upfront cost[591,592].

Recall, the upfront cost is already very high for BEVs because of the costly battery system. Major advances will be required in BEV technology to increase the travel range and concurrently lower the upfront cost.

Thus, massive advances are required in charging technology as well as vehicle technology. Such advances will require many years.

Summary: BEVs have major convenience challenges. For example, public charging stations are currently dominated by charging stations that require a few to several hours for providing a 200-mile charge. The fast-charging stations, which require 15 to 60 minutes, are substantially more expensive. Moreover, even the fast-charging stations are at-least a factor of three slower than conventional gas fueling stations. Consequently, advances will be required in charging technology for increasing charging speed and simultaneously reducing costs. BEVs also have a shorter travel range compared to conventional vehicles. Comparable travel range can be obtained for BEVs but at a substantially increased upfront and total lifetime cost. Major advances will be required in BEV technology to increase the travel range and substantially lower costs. Overall, massive advances will be required in charging technology as well as BEV technology for alleviating the convenience challenges. Such technology advances will require many years.

The Debate Around Direct CO_2 Capture Technology

Can CO_2 capture from air become a wide-scale solution in the foreseeable future?

The technology is very appealing because it can directly reduce the CO_2 content of the atmosphere. But there are many challenges.

Direct CO_2 capture from air has not been implemented on a practically relevant scale. Therefore, accurate cost estimates for large-scale applications are not available. However, CO_2 capture from fossil fuel power plants can be used as a reference case to discuss the direct CO_2 capture cost on a relative basis.

In case of CO_2 capture from power plants, the concentration of CO_2 in the byproduct stream is typically above 5%[593]. In contrast, the CO_2 content in the atmosphere is 0.04%[594]. Thus, the CO_2 in the atmosphere is over 100 times more dilute than in the byproducts from power plants.[595]

CO_2 capture from power plants has high associated costs[596]. Direct CO_2 capture from air is far more challenging because atmospheric CO_2 is extremely dilute. Thus, it follows that direct CO_2 capture from air will have much higher costs than CO_2 capture from power plants.

Also, it is energetically inefficient to convert CO_2 into useful products. Hence, no significant cost benefits are expected from converting the captured CO_2 into useful products such as chemicals[597].

Direct CO_2 capture from air is severely challenged by laws of nature because of the extreme diluteness of CO_2 in the atmosphere. Hence, cost breakthrough claims will need to be carefully evaluated. Such an evaluation requires capital cost and operating data on several large-scale operational units. The data is not expected for at-least ten years.

The deployment of direct CO_2 capture from air will be very limited in the next couple of decades considering the very high anticipated costs and high technology risk.

However, nature-based solutions such as using trees for direct capture of CO_2 from air can be excellent solutions in the near term, when deployed strategically.

Summary: The direct CO_2 capture from air is severely challenged by laws of nature. The concentration of CO_2 in the atmosphere is 100 to 300 times lower than the concentration of CO_2 in the byproduct stream of power plants. Despite the higher concentration, CO_2 capture from power plants has been found to be challenging and expensive. The costs and challenges for direct CO_2 capture from air technology are, therefore, expected to be very high. Cost breakthrough claims will need carefully evaluation based on data from several large-scale units. Such data is not expected for at-least ten years.

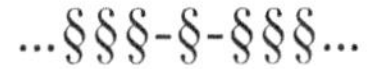

9. Key Issues Around Climate and Energy

"Science is organized common sense where many a beautiful theory was killed by ugly facts"–***Thomas Huxley***

"Look deep into nature and then you will understand everything better"–***Albert Einstein***

This chapter addresses the widely misunderstood issues around human-caused climate change and energy transition. A realistic understanding of these issues is essential for productive discussions about how to best address climate change.

Human-Caused Climate Change: Key Issues

Should we ignore the conclusions from climate research because certain past claims have been unreliable?

Over the decades, media and certain scientists have greatly exaggerated the speed and magnitude of the impact from climate change. For example, the renowned physicist John Holdren proposed in the 1980s that human-caused climate change could kill a billion people because of famine by the year 2020[598]. Such proposals have no credibility because they are purely speculative, i.e., they are not based on robust science[599]. The global scientific community does not support such proposals.

Most climate experts agree that human-caused climate change is a serious problem that requires urgent attention[600,601]. Major scientific organizations around the globe agree with these conclusions because they are based on extensive research.

Wild exaggerations by media or certain scientists cannot change the validity of these conclusions

We will consider an analogy. The medical community has concluded that smoking tobacco is dangerous to human health based on extensive research[602,603]. Wild exaggerations by a few doctors–made in the past, present or future–do not decrease the trustworthiness of the conclusion.

Summary: Wild speculations by certain scientists and media should not be used an excuse to ignore critical findings from the global climate experts. The findings of the climate experts are based on extensive research. Consequently, major scientific organizations agree that human-caused climate change is a serious problem that requires urgent attention. Wild exaggerations by media or certain scientists do not change the validity of these conclusions.

Should we ignore climate change since deaths from climate disasters have decreased drastically?

The World Meteorological Organization recently published a report documenting decade-by-decade trends in weather, climate and water-related disasters from 1970 to 2019[604]. Per the report, death toll fell from over 500,000 deaths in the 1970s to less than 200,000 deaths in the 2010s. The greenhouse gas emissions increased from 25 billion tons to 52 billion tons during the same period[605]. This data appears to be contradictory at first glance, but it is not.

The number of disasters increased over four-fold, from 711 recorded disasters in the 1970s to 3165 disasters in the 2010s[606,607]. Deaths have markedly decreased because of the advances in multi-hazard early warning systems and superior medical access. The early warning systems have allowed decision makers, communities, and individuals to be better prepared and have saved many lives.

Deaths are not the only metric of importance. These disasters have wreaked havoc on millions of lives by causing loss of homes, livelihood, and property[608]. The economic losses have been enormous, for example, the reported losses from the disasters amounted to 1.4 trillion dollars from 2010 to 2019[609].

The recent IPCC report concludes that the number and severity of disasters will increase manyfold for the scenarios that do not mitigate climate change[610]. Over time, this will cause major hardships to a substantial fraction of the global population.

Thus, climate change impacts are already significant and are expected to become more worrisome with time. Apart from direct impacts–such as deaths, loss of homes, livelihood, and property–several indirect impacts are also expected. For example, climate-induced water scarcity could force massive migrations[611].

Summary: Although the death rates have decreased drastically over the decades, the number of disasters has increased by over a factor of four. Death rates have decreased because of early warning systems that have allowed communities to take life-saving action such as timely evacuation. The disasters have wreaked havoc on millions of lives by causing loss of homes, livelihood, and property. According to climate experts, the last few decades have provided robust data to state with high confidence that the number of disasters will increase rapidly if climate change is not mitigated.

How can climate change be a serious problem when it is yet to be understood perfectly?

The lack of perfect understanding about climate change science is discussed by Steve Koonin in his book "Unsettled"[612]. Dr. Koonin is a respected theoretical physicist, who served as the Undersecretary for science under the Obama administration.

He discusses several issues in the book. Specifically, he highlights the uncertainties related to climate change studies. He also discusses the exaggerations about human-caused climate change by media. He concludes that climate change is not a serious problem.

Is this conclusion consistent with the collective body of scientific literature?

Climate change science is an extremely complex area with intricate interactions between the atmosphere, biosphere, cryosphere, hydrosphere, and land surface. Correspondingly, uncertainties are expected and will never be eliminated. However, scientific studies can reduce the uncertainties.

Climate researchers have been undertaking massive efforts over the past decades to decrease the uncertainties[613]. Many important issues around climate change are now understood with high to very high confidence because of these efforts. A few highlights from the most recent IPCC flagship report are included below[614]:

- There is very high confidence that the increases in carbon-dioxide and methane concentrations since 1750 far exceed the natural changes over at-least the past 800,000 years.
- It is unequivocal that human influence has warmed the atmosphere, ocean and land. There is high confidence that global surface temperature has increased faster since 1970 than in any 50-year period over at-least the last 2000 years.
- There is high confidence that global mean sea level has risen faster since 1900 than over any preceding century in at-least the last 3000 years. There is over 90% certainty that human influence is responsible for the increase in sea levels.
- There is over 99% certainty that hot extremes including heat waves have become more frequent and more intense across most land regions since the 1950s, while cold extremes including cold waves have become less frequent and less severe.

- There is over 99% certainty that human-caused emissions are increasing the acidity of the global surface open oceans.
- There is over 90% certainty that human influence is responsible for the global retreat of glaciers since the 1990s.

The global climate science community has high confidence or certainty about the above impacts because adequate data is available. Moreover, there is a potential for several other severe climate change impacts. These impacts are currently understood at low-to-moderate confidence because of the limited data.

Notably, the impacts that are understood with high confidence are by themselves adequate to demonstrate that human-caused climate change is a serious problem.

Think about stock market investments. Despite significant uncertainty, the stock market is an important investment strategy amongst the investment community. Why? Because despite the uncertainty or lack of perfect understanding, the investment community is confident that investing in stocks is essential for improving the long-term outcome of their investments.

Summary: The current understanding of climate change is less than perfect. For example, several impacts from climate change are not yet understood with high confidence because of limited data. However, many important impacts from climate change are very well-understood because adequate data is available related to those impacts. The impacts which are understood with high confidence are sufficient to demonstrate that human-caused climate change is a serious problem.

What is the consensus from climate experts about the future impacts of climate change?

Most climate experts agree with the following[615,616]:

- The impact of human-caused climate change will increase markedly with increasing global warming. Limiting the amount of warming requires urgent global actions.
- Severity of the impact can be decreased via urgent investments in proactive climate adaptation.
- Developing countries will be impacted more severely than the developed countries because of inferior availability of resources and more susceptible locations.

While the scientific community agrees that the severity of climate change impact will markedly increase with increasing temperature, there is no consensus for 2.5°C or an even 4°C temperature rise being catastrophic for human existence. In other words, the end of the world or doomsday discussions related to climate change are not credible.

Summary: The consensus amongst climate experts is that the severity of future climate change impact will depend on the trajectory of global greenhouse gas emissions. Also, climate experts agree that low-income countries will be impacted more severely because of resource and location constraints and that large investments in climate adaptation will be crucial for decreasing impacts.

What is the estimated impact of climate change on the economy?

The impact on the economy is usually discussed in terms of the impact on the Gross Domestic Product (GDP). Recall, GDP is the monetary value of all goods and services made within a country or region during a specific period.

For example, the recent COVID-19 pandemic had a large impact on the global economy. The world GDP decreased by 3.3% in 2020[617]. Without the COVID-19 pandemic, the world GDP was estimated to grow by about 3% in 2020[618]. Thus, the COVID-19 pandemic caused a loss of over six percentage points to the global economy in 2020[619].

The U.S. Congressional Budget Office (CBO) has recently provided an estimate for the impact of climate change on the U.S. economy[620]. The estimate was based on the average of moderate and high greenhouse gas emission scenarios.

The average annual GDP growth in the U.S. has been about 3% since 1960[621]. The CBO study estimated that climate change will reduce the average U.S. annual growth of GDP by 0.03 percentage points from 2020 to 2050 relative to the climatic conditions that prevailed at the end of the twentieth century[622]. For example, there would be a 2.97% annual GDP growth instead of a 3% annual GDP growth due to the climate change impact.

The 0.03 percentage point reduction in annual GDP growth rate would accumulate to a 1% reduction in the U.S. GDP over the projected period of 30 years. Thus, the U.S. GDP would be about 1% smaller in 2050 compared to what it would be without climate change according to the CBO estimate.

A recent study by the insurance company Swiss Re was in good agreement with the CBO study when the current understanding about climate change impact was included[623]. However, when the Swiss Re study applied large arbitrary factors to account for potential unknown climate change issues[624], there was a large divergence from the CBO study.

Based on the currently known climate change impacts, the Swiss Re study estimated that the U.S. GDP would be about 1% smaller in 2050 compared to what it would be without climate change. However, by including an arbitrary tenfold multiplication factor for unknown impacts, the study predicted that the U.S. GDP would be about 8% smaller in 2050.

Based on the known climate change impacts, the Swiss Re study estimated that the world GDP would be about 2% smaller in 2050 compared to what it would be without climate change[625]. But when they included an arbitrary tenfold factor to account for potential unknowns, the impact on the world GDP in 2050 jumped to a 18% loss for their worst-case scenario.

The potential unknowns could increase the economic impact from climate change. But it is scientifically impossible to make robust estimations for the impact from the potential unknowns. Consequently, the most reasonable approach is to only use the known impacts for the estimation and acknowledge the uncertainty.

The Network for Greening the Financial System (NGFS) has recently estimated that the global GDP would be smaller by 3% in 2050 if current energy policies are followed as opposed to net zero 2050 policies[626,627].

Summary: Studies have shown that the GDP will grow at a slower rate because of climate change impacts. A recent U.S. CBO study recently estimated that the annual growth of the U.S. GDP would be smaller by 0.03 percentage points until 2050. This equates to the U.S. GDP being smaller by a total of 1% in 2050 because of climate change. According to a recent study by the insurance company Swiss Re, the global GDP would be smaller by 2% in 2050 when the known climate impacts are included. But when the study included an arbitrary tenfold factor to account for potential unknowns, the impact on the global GDP jumped to a 18% loss in 2050. The latter estimation is not credible because it is scientifically impossible to make robust estimations for the impact from the potential unknowns[628]. Recently, NGFS estimated that the global GDP would be 3% smaller in 2050 for current policies as opposed to net zero 2050 policies.

Is climate change mitigation the most important issue for humans currently?

Human-caused climate change is a serious problem and requires urgent attention from the global population. But is it the most important issue for humans currently?

The scientific community or think tanks cannot answer this question. It depends on the global population. Specifically, it depends on whether the overwhelming majority believes that climate change is their most important issue.

How to determine this? By considering key facts about the global population.

First, we will consider the population that belongs to the economically distressed category. For this discussion, the economically distressed category is the population living under $10/day[629].

People in the economically distressed category have extremely small budgets. Therefore, they are forced to compromise on one or more critical needs every day. They must choose between adequate nutrition, energy, medical, housing quality, or education[630,631]. Each day is a struggle for meeting basic needs.

Climate change is expected to have the worst impact on the economically distressed. Hence, their future state is expected to worsen if climate change is not mitigated.

Does this mean they will support climate change mitigation as the most important issue? Not likely!

Why? Because the current state is far more important than the future state for people who are suffering. For example, if a person is in severe pain from an existing disease, that person is unlikely to be overly worried about the possibility of getting cancer in the distant future. From that person's viewpoint, getting relief from existing pain is more important.

Thus, the economically distressed do not have the luxury of worrying about future issues from worsening climate change. Their most pressing problem is meeting basic needs.

What fraction of the global population belongs to the economically distressed category?

Based on data from the World Bank, about 50% of the global population or about 3.5 billion people are estimated to live under $10 per day[632,633]. The people in this category are unlikely to consider climate change as their most important problem.

In good agreement, a recent global survey by the United Nations Development program showed a 55% public support for climate policies, from high income countries, and only 36% public support from the least developed countries[634,635]. Globally, there was only 38% public support for urgently doing everything necessary for climate change mitigation[636].

Thus, the major fraction of the global population currently does not consider climate change as the most important issue[637,638]. The affluent–such as those with easy access to adequate nutrition, energy, medical, comfortable housing, and education–are far more likely to consider climate change to be the most important issue[639,640].

But even amongst the affluent, a sizeable fraction is not expected to consider climate change as the most important issue. Why? Because some other critical issue could be more important for them. Examples of other critical issues include national security, energy security, and threats from nuclear war, pandemics, autocracy and artificial intelligence.

Summary: The answer to the question depends on who is being asked. About 50% of the global population lives under $10/day. This economically distressed category does not have access to basic needs such as adequate nutrition, energy, medical treatments, housing, and education. They are currently suffering. Hence, they do not have the luxury to worry about future impacts from worsening climate change. On the other hand, climate change is the most important problem for people who are affluent and are extremely passionate about the environment. But even amongst the affluent many are likely to give higher importance to national security, energy security, or threats from nuclear war, pandemics, autocracy and artificial intelligence.

How long did it take to definitively understand that CO_2 from fossil fuels was a serious problem?

Climate change is the most important environmental problem from fossil fuel technologies[641]. But the scientific community required a very long time to definitively recognize this.

The historical timeline is discussed below.

Large-scale combustion of coal started in the mid eighteenth century. The CO_2 impact from coal combustion was first studied over a century and a half later by the Swedish scientist Svante Arrhenius. Towards the end of the nineteenth century, he proposed that CO_2 released from coal combustion would increase earth's temperature[642]. By then, 45 billion tons of CO_2 had been released from fossil fuel combustion[643,644].

Arrhenius' research did not draw attention from the scientific community. Thus, the scientific community had no concern about severe climate impact at the end of the nineteenth century.

In the 1930s, the English engineer Guy Callendar proposed that earth was already warming because of the CO_2 emissions from fossil fuel combustion[645]. Notably, he believed that this effect would be beneficial in the long term[646]. By 1935, the global CO_2 from fossil fuel combustion had crossed an imposing 150 billion tons.

Callendar's work prompted further studies in the 1950s and 1960s. These studies generally agreed that warming was occurring because of the CO_2 emissions from human activity. Based on these studies, the scientific community for the first time recognized the potential for a severe impact from fossil fuels. But they acknowledged the large quantitative uncertainties in predicting the effects. This is documented in key historical reports.

In 1965, a comprehensive report was published by the United States President's Science Advisory Committee titled "Restoring the quality of our environment", wherein several important recommendations were made to ensure a healthier environment[647]. The final recommendations in the report are indicative about the prevailing scientific views.

The report recommended several concrete actions to decrease air and water pollutants. However, the scientists made only one significant recommendation related to CO_2. It was to continue the precise measurements of CO_2 and temperature at different heights in the stratosphere. Clearly, the seriousness of the problem was not adequately understood back then[648]. By the year 1970, the cumulative global amount of CO_2 released from fossil fuel combustion had exceeded 400 billion tons[649].

In the 1970s, several studies expressed concern about the long-term impact of warming from greenhouse gases. However, some studies also raised an alarm about potential global cooling because of the dust and smog produced from human activities. Evidently, the scientific community lacked adequate understanding of the climate change problem. This substantial gap in knowledge was highlighted in the special report about Climate Change by the U.S. National Academy of Science in 1975[650]. For reference, a defining statement from the report is provided below.

"Our knowledge of the mechanisms of climate change is at least as fragmentary as our data. Not only are the basic scientific questions largely unanswered, but in many cases we do not yet know enough to pose the key questions."

By the end of 1980, the global CO_2 emissions from human activity had reached 600 billion tons[651]. Studies related to the warming from greenhouse gases received increased attention in the 1980s.

Global climate experts became very concerned about human-caused climate change because of these studies. The Intergovernmental Panel on Climate Change (IPCC) was established in 1988 to guide global Governments on this issue[652].

IPCC published their first flagship report in the early 1990s[653]. The report concluded that CO_2 was the main contributor to warming. It also concluded that earth's mean temperature and sea levels would increase significantly by the year 2100, if the use of fossil fuels was not limited.

However, their judgment section revealed that a *definitive understanding* had not yet been achieved. For a definitive understanding, the CO_2 effect on temperature needed to be clearly distinguished from natural climate variability. The exact statements from the report are provided below as evidence for the absence of such knowledge at that time.

- *"Global mean surface air temperature has increased by 0.3°C to 0.6°C over the last 100 years, with the five global average warmest years being in the 1980s."*
- *"The size of the warming is broadly consistent with predictions of climate models, but it is also of the same magnitude as natural climate variability."*
- *"The unequivocal detection of the enhanced greenhouse effect from observations is not likely for a decade or more"*.

By 1990, the global cumulative CO_2 emissions had exceeded 750 billion tons. As predicted in the report, it took until the end of the century to unequivocally detect the CO_2 impact. Thus, the CO_2 impact from fossil fuels was definitively understood only at the end of the twentieth century[654]. By then, fossil fuels had been in use for two and a half centuries and had generated 1000 billion tons of CO_2.

Summary: Past publications inform us about the evolution of the understanding about the topic. Energy production from fossil fuels grew rapidly from the mid 1700s. But the scientific community had no concern about the growing CO_2 emissions for the next two hundred years. In fact, the CO_2 produced from fossil fuels was speculated to be beneficial in the long term. Notably, fossil fuel combustion had generated over 200 billion tons of CO_2 by then. Although there was concern about global warming in the 1960s and the 1970s, the scientific community acknowledged their inadequate understanding about climate change. The challenge faced by the scientific community is further recognized from the first IPCC report published in the early 1990s. The judgment section in the report acknowledged that the observed warming was too low to be conclusively attributed to CO_2 emissions. By then, 700 billion metric tons of CO_2 had been released. The impact of the CO_2 emissions was only definitively understood at the end of the twentieth century–two and a half centuries after the first large scale use of fossil fuels and after 1000 billion tons of CO_2 had been released.

Why did the scientific community take so long to definitively understand the fossil CO_2 impact?

It took a very long time to definitively understand the most important environmental problem from fossil fuel technologies. We must understand the underlying reason because this will help in anticipating future problems with the emerging energy technologies.

First, we need to ask: What is the typical process to definitively understand the environmental problem resulting from a technology?

It is a two-step process.

- The first step establishes a scientific basis for the possibility of a severe environmental problem. A scientific basis requires substantial evidence. Without substantial evidence, it is a theory not a scientific basis. Substantial evidence requires a wide-scale deployment of the technology. Why? Because when the technology is deployed below a certain minimum level, the environmental impact from the technology is too small for clear detection.
- The second step provides conclusive evidence about the severity of the problem based on direct observations. Expectedly, such evidence requires an even wider-scale deployment of the technology.

Thus, two minimum levels of deployment must be reached to definitively understand the environmental problem from a technology. The first level needs to be reached to establish a scientific basis. The second level needs to be reached to secure conclusive evidence.

The deployment of fossil fuel technologies did not reach the first level until the mid-twentieth century. How do we know that? Because there was no concern about climate impact until then[655].

A scientific basis for the possibility of severe climate impact from fossil fuel technologies was proposed only in the 1960s[656]. By the mid-1980s, the scientific basis became robust because of the additional data from higher deployment levels.

But conclusive evidence was still missing. Recall, the IPCC report published in the early 1990s stated that the size of the additional warming was the same magnitude as natural climate variability, and that a definite detection of the enhanced greenhouse gas effect was not possible at that time[657].

Another decade was required for the definitive confirmation from actual observations. Why? Because the large additional deployment of fossil fuel technologies in that decade caused a large increase in greenhouse gas emissions. For example, the cumulative emissions of CO_2 increased from 800 billion tons in 1990 to 1000 billion tons in 2000[658]. The rise in emissions finally increased the temperature to a level that allowed the impact from fossil fuels to be unmistakably distinguished from natural variability[659]. Also, the climate change impacts such as rising sea levels and loss of snow-cover became significantly more noticeable.

Summary: Two steps are required to develop a definitive understanding about the environmental impact from a technology. The first step establishes a scientific basis for the possibility of a severe environmental problem. In the second step, conclusive evidence is secured from direct observations. Each step requires a certain minimum level of technology deployment to obtain adequate data. In case of fossil fuel technologies, it took two and a half centuries to reach the minimum deployment level that was required to obtain conclusive data. Specifically, the temperature rise from the CO_2 emissions was large enough to definitively distinguish the fossil fuels impact from natural causes only at the end of the twentieth century.

Will solar and wind not have any severe environmental impact when deployed at a wide-scale?

To address this question, we will consider the lessons from the historical data on fossil fuel technologies.

The most important environmental problem from fossil fuel technologies–its impact on the climate–was definitively understood only after a very wide-scale deployment of fossil fuel technologies. At low deployment levels of fossil fuel technologies, for example in the 1940s, there was no concern about the climate impact from fossil fuels[660]. In fact, the pioneering researcher Guy Callendar–who published the ground-breaking paper on human-caused CO_2 emissions and global warming in the late 1930s–believed that the warming effect from fossil fuels would be beneficial in the long term[661,662].

Recall, two minimum levels of deployment must be reached to definitively understand the environmental problem from a technology. The first level needs to be reached to establish a scientific basis. The second level needs to be reached to secure conclusive evidence.

A scientific basis for the possibility of a severe climate impact from fossil fuel technologies was established only in the 1960s[663].

Why was the potential for severe climate impact not recognized earlier? Because the required information was not available at the lower fossil fuel deployment levels. Specifically, below the first minimum level of deployment of fossil fuel technologies, the climate change impact from the CO_2 emissions was not evident and therefore could not be considered as reasonable evidence. Thus, before the 1960s, the required information was not available to establish a scientific basis.

Why is this important? Because renewable technologies such as solar and wind are currently deployed at extremely low levels. For reference, solar and wind energy combined in 2020 produced only 15% of the energy produced from fossil fuels way back in 1940[664,665]. Recall, in 1940, the scientific community had no concern about the climate impact from fossil fuels. In other words, there was no concern about the climate impact from fossil fuels even when deployed at six-fold higher levels than the current levels of solar and wind[666].

Thus, renewable technologies are currently far below the minimum deployment level that would be required to establish a scientific basis for the possibility of a severe future problem. This learning from fossil fuel technologies has enormous implications. The learning informs that the most important environmental impacts from these technologies will only be recognized after they are widely deployed.

Specifically, these technologies would need to produce at-least as much energy as was produced from fossil fuel technologies in the 1940s. Why? Because the scientific basis for the possibility of severe climate impact from fossil fuels was established only in the 1960s.

Next, we will consider the lessons from basic scientific principles.

Solar, wind and fossil fuel technologies have the same primary energy source, i.e., solar energy. Also, these technologies have an identical final product, which is 24X7 electricity. Thus, solar, wind and fossil fuel power technologies have an identical primary energy source and final point. Correspondingly, the magnitude of support or helping hand from nature determines the amount of human activity and resources required for producing 24X7 electricity.

In case of fossil fuels, the solar energy captured in ancient plants and organisms has been concentrated in large deposits around the globe. Nature has undertaken this transformation process via application of heat and pressure on the ancient plants and organisms in the earth's crust over millions of years[667]. Nature, thus, has provided a large advantage for fossil fuels, i.e., nature has shortened the path for 24X7 electricity generation for fossil fuel technologies by eliminating the extreme diluteness and intermittency challenges associated with the primary energy source (solar energy).

Solar and wind power do not have such a helping hand from nature, i.e., their critical challenges such as intermittency and extremely diluteness are not resolved by nature. Consequently, solar and wind technologies require substantially higher human activity and resources for producing 24X7 electricity compared to fossil fuel technologies[668].

Large requirements of human activity and resources equals high environmental impact[669,670]. Considering that solar power and wind power have much higher requirements of human activity and resources compared to fossil fuel technologies, these technologies are expected to have a high potential for some form of severe environmental impact, when deployed at a scale comparable to fossil fuel technologies[671]. The severe impact could be in any form such as freshwater eutrophication, and/or human toxicity, and/or wildlife endangerment, and/or water resource depletion, and/or climate impact, and/or minerals depletion and/or land use.

For the power sector, examples of human activity include mining, transportation of materials, pretreatment of materials, manufacturing, installation and land preparation. Resources include land, water, minerals, energy and materials.

An example for a solar power plant, which is comparable in capacity to an average natural gas power plant, is discussed below.

The 648 MW capacity Kamuthi solar power plant project was completed in 2016 in Tamil Nadu, India. The solar power plant consists of 2,500,000 solar modules, 380,000 foundations, 30,000 tons of structures, 576 inverters, 154 transformers, and 6000 kms of cables[672,673]. The Kamuthi power plant is spread across 2500 acres, i.e., land equivalent to about 950 Olympic sized soccer fields. For reference, a similar capacity natural gas power plant would require 25 acres of land and a small fraction of materials[674].

Human activity was required for preparing the land, mining, manufacturing and installing the various components of the solar power plant. Resources include the land, energy and the materials required for the different components. Also, large amounts of water are required to regularly clean the two and a half million solar panels[675]. Thus, a solar power plant requires a large amount of human activity and resources.

But this does not provide the complete picture about the human activity and resource requirements for solar power.

Unlike fossil fuel plants, solar power plants cannot produce 24X7 electricity on a standalone basis. To do so, the power plant requires coupling with an energy storage technology such as batteries. The energy storage capacity will have to be very large to ensure 24X7 electricity because of the daily and seasonal variation in sunlight availability[676]. A large amount of human activity and resources will also be required for the energy storage plant[677]. At a wide-scale deployment, the total human activity requirements and resources from the solar power, wind power and energy storage will become enormous and will very likely result in some form of severe environmental impact.

Summary: The learnings from historical data inform that severe environmental impact(s) from a technology will only be known after it is widely deployed. Solar and wind power are currently deployed at very low levels. Unlike fossil fuels, solar and wind power do not have a massive helping hand from nature. Correspondingly, they require far more human activity and resources to produce 24X7 electricity. High requirements of human activity and resources equals high potential for severe environmental impact. Therefore, solar and wind power have a high possibility for some form of severe environmental impact, when deployed at a scale comparable to fossil fuel technologies.

Is it appropriate for media to make sensational claims based on isolated peer reviewed articles?

Over two million research articles are published each year[678]. A peer review process is used to regulate the quality of the information. Two or more experts review the research article during a typical peer review process. Scientific journals accept the article for publication only if the reviewers provide favorable comments. This is an efficient approach for the quality control of research articles.

However, the peer review process has certain limitations.

The peer review process is unpaid and voluntary. The peer reviewers need to take time from their daily job functions to help with the peer review process. Many researchers review several articles each year because of the poor ratio between reviewers and the number of submitted articles. Therefore, reviewers can dedicate only a short time to evaluate the article and provide comments.

Reviewers use this limited time to check if the overall approach is reasonable and the conclusions are consistent with the presented data.

But the peer review process cannot ensure the accuracy of a research article. Why? Because an extensive study of the data collection method, data quality, and data analysis is required to check accuracy. This would require several weeks of focused work and is not practical.

Despite the best intentions of the scientific community, every research article has potential for inaccuracy related to data collection, data quality or data analysis. Ultra-complex fields such as climate science have more potential for inaccuracies. Therefore, conclusions from an individual or isolated research paper are not appropriate for making a substantial claim[679].

We will discuss an example. According to recent discussions in the prestigious journals Nature and Science, certain isolated articles have been reporting exaggerated climate impacts because of their use of inferior climate models[680,681].

Fortunately, this issue has been correctly dealt with in the recent IPCC report, i.e., accurate climate models have been used for discussing future climate impacts in the IPCC report[682]. The IPCC report is substantially more credible because the discussions in the report are drawn from the efforts from multiple global climate scientists with a wide range of expertise.

A sensational claim in any complex field requires consistent conclusions from a series of research articles, i.e., it should be consistent with efforts from many global experts[683]. This is especially true for ultra-complex and empirical fields such as climate science.

Summary: Despite the good intentions, all research articles have a potential for inaccuracy related to data collection methods, data quality or data analysis. The potential for inaccuracy is higher for ultra-complex fields such as climate science. The peer review process used to regulate the quality of the articles has certain limitations. A typical peer review process only checks whether the approach used in the article is reasonable and whether the conclusions are consistent with the presented data. The peer review process cannot ensure the accuracy of a research article because that would require several weeks to months. Media should not rely on the conclusions from isolated articles to make sensational claims about climate issues because of the significant potential for inaccuracy. Consistent conclusions from multiple research articles are required to make substantial claims about climate issues.

Energy Transition: Key Issues

What sectors are the major contributors to global greenhouse gas emissions?

Annual greenhouse gas emissions have exceeded 50 billion tons in recent years[684].

The energy sector is the largest contributor with a 76% share of the global emissions[685,686,687]. The share of emissions from agriculture, land use change and forestry combined is 15%, while that from direct industrial processes and waste is 6% and 3%, respectively[688].

Sub-sector level data provides the maximum potential for emission reductions from each individual category. Two examples are discussed.

Coal power plants generate 19% of the global greenhouse gas emissions, while light duty vehicles generate 7%. Thus, replacing coal power plants with low-carbon power technologies will have a much larger impact compared to replacing conventional light duty vehicles with electric vehicles. Such information is essential for a robust path forward discussion.

Summary: Energy is the largest contributor with a 76% share of the global greenhouse gas emissions. Agriculture, land use change and forestry combined have a 15% share, while direct industrial processes and waste combined have a 9% share of the global greenhouse gas emissions. Subsector level data provides information about the relative importance of individual categories in reducing greenhouse gases.

What countries are the major contributors to greenhouse gas emissions?

Two separate metrics are required to discuss the relative contribution to greenhouse gas emissions from different countries.

The first metric is the cumulative contribution to greenhouse gas emissions. This metric provides relative information about the contribution from the different countries to human-caused climate change.

The cumulative contribution from a country is estimated by including all its greenhouse gas emissions from the earliest year for which data is available. CO_2 is used as a proxy for the greenhouse gases because it is the major component and historical CO_2 data is available.

The top emitters based on the cumulative contribution to CO_2 emissions are United States (24% of global share), EU-28 countries (22% of global share), China (14% of global share), Russia (7% of global share), and Japan (4% of global share)[689]. These top emitters have a 71% share in terms of cumulative contributions to the global CO_2 emissions.

The second metric is the current contribution to greenhouse gas emissions. This metric provides information on the extent of greenhouse gas reduction that would be required by each country to meet net zero goals[690].

The top emitters based on the current contribution to greenhouse gas emissions are China (27% of global share), United States (13% of global share), EU-28 countries (8% of global share), India (7% of global share) and Russia (5% of global share)[691]. These top emitters have a 60% share in terms of the current contributions to global greenhouse gas emissions.

The relative contribution has changed over the decades. For example, the emissions from China and India have increased markedly, while that from the United States and EU-28 countries have decreased over the past couple of decades (**Figure 9.1**).

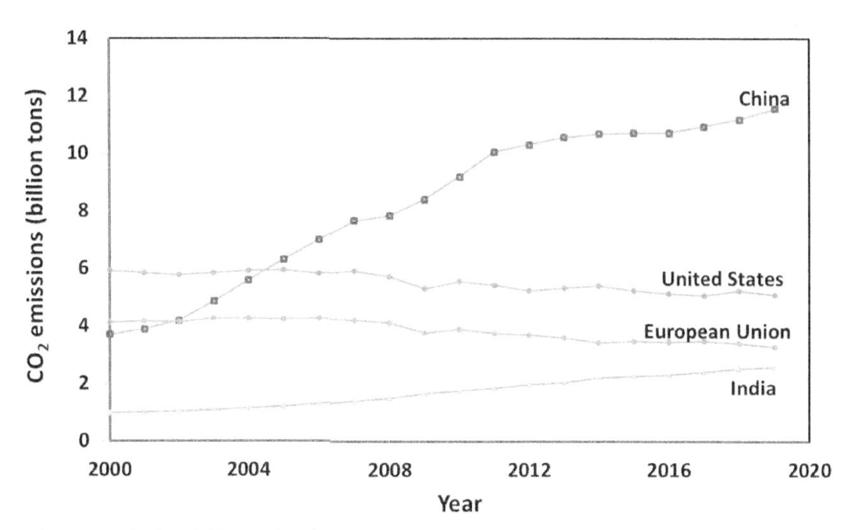

Figure 9.1: Historical CO_2 emission trends from representative countries. Source: Netherlands Environmental Assessment Agency.

Summary: A discussion about the relative contribution to greenhouse gas emissions from countries requires two metrics: a) cumulative contribution over the decades and b) current contribution. The top emitters based on cumulative contribution to CO_2 emissions are United States (24%), EU-28 countries (22%), China (14%), Russia (7%), and Japan (4%). This metric provides relative information about the contribution of the country to human-caused climate change. The top emitters based on the current contribution to greenhouse gas emissions are China (27%), United States (13%), EU-28 countries (8%), India (7%) and Russia (5%). This metric provides information on the extent of greenhouse gas reduction that would be required by the country to meet net zero goals.

What are the key challenges related to electrification?

Electrification is the process of replacing technologies that use fossil fuels with technologies that use electricity. Examples are replacement of conventional vehicles with BEVs and replacement of gas furnaces with heat pumps.

Electrification can drastically reduce greenhouse gases when the electricity is generated from low-carbon technologies. Hence, it is a major pathway for the low-carbon energy transition.

But the benefit from electrification is limited if the electrical grid does not have a low carbon intensity. For reference, BEVs reduce greenhouse gases by less than 25% in regions that have an electrical grid with high carbon intensity.

Electrification technologies can be classified into two broad types: established and under development[692]. Examples of established technologies include battery electric cars and heat pumps.

But even the established electrification technologies suffer from either high upfront costs, and/or high total lifetime costs and/or substantial inconvenience. Large advances are required for these technologies to be competitive to fossil fuel technologies in terms of costs and convenience. Recall the previous discussion on battery electric light duty vehicles.

Examples of electrification technologies that are under development include battery technologies for road freight, shipping and aviation and technologies that can provide the high temperatures required for certain industrial processes[693,694].

A substantial cost and performance risk is associated with the electrification technologies under development. Why? Because these technologies are focused on the difficult-to-decarbonize sectors. Recall the earlier discussion about the severe challenges associated with battery applications for road freight, shipping and aviation.

Electrification cannot eliminate greenhouse gases from processes that naturally produce CO_2 such as the cement industry[695]. Consequently, other technologies are also being developed to address the difficult-to-decarbonize sectors. Examples of such technologies are carbon capture and storage, green hydrogen, and biofuels.

Summary: Electrification can drastically reduce greenhouse gases when the electricity is generated from low-carbon technologies. Hence, electrification is a major pathway for a low-carbon energy transition. But electrification is inefficient for reducing greenhouse gases if the electrical grid does not have a low carbon intensity. Moreover, even the established electrification technologies–such as battery electric cars and heat pumps–currently suffer from either high upfront costs, and/or high total lifetime costs and/or inconvenience challenges. Major advances are required for these technologies to be competitive with fossil fuel technologies. Electrification technologies are also being developed to address the difficult-to-decarbonize sectors. Such technologies have a high viability risk because of huge cost and performance challenges.

What is the cost for a global low carbon transition?

The global consulting firm McKinsey & Company recently provided a cost estimate for a global low carbon transition[696,697]. Their analysis revealed an additional cost of 3.5 trillion dollars per year to the global society from 2021 to 2050.

How much is 3.5 trillion dollars per year? The figure is approximately equivalent to half of annual global corporate profits or one quarter of the annual global tax revenue[698].

Their cost estimate is based on the physical assets required for the low-carbon transition. The estimate does not consider the supporting costs that would also be required. Examples of excluded costs are a) retraining of the workforce, b) compensation for stranded assets, c) loss of value pools in parts of the economy, and d) redundancy in energy systems that would be required to avoid supply volatility during the transition.

These excluded costs are expected to be enormous considering that a very large workforce will need to be retrained, vast number of assets will be stranded, and substantial redundancy in energy systems will be required[699]. Thus, the estimated costs are conservative, i.e., they are low-end costs.

How much retraining will be required? The study estimated that the energy transition will lead to a gain of 200 million jobs and loss of 185 million jobs by 2050. Thus, a massive retraining will be required to balance the work force.

Some climate activists want the transition to occur more quickly. For reference, if the transition was to occur over a ten-year period instead of thirty years, the annual additional costs to the global society would be over 10 trillion dollars per year.

Summary: The global consulting firm McKinsey & Company recently estimated that a global low carbon transition would add a cost of 3.5 trillion dollars per year from 2021 to 2050. This additional cost is roughly equivalent to half of annual global corporate profits. Moreover, the estimates are on the low end because the supporting costs such as retraining of the workforce, compensation for stranded assets, redundancy in energy systems and loss of value pools in parts of the economy are not included.

What are key concerns about the critical minerals required for a low-carbon energy transition?

Critical minerals are those that are required for important applications and are at risk for supply disruption[700]. Example applications are high technology devices, defense applications and energy technologies.

Low-carbon technologies require a wide range of critical minerals. For example, batteries require cobalt, lithium, nickel, manganese and graphite, whereas wind turbines and electric vehicles require rare earth elements[701].

Abundant access to critical minerals is crucial for the popular low-carbon energy technologies. Why? Because these technologies require several-fold larger amounts of critical minerals compared to conventional technologies[702].

Currently, adequate access to critical minerals is not an issue because the deployment levels of low-carbon technologies are very small. For example, wind and solar combined provide less than 5% of the total global energy and electric cars represent 1% of the total global car stock[703,704].

But a low-carbon energy transition will cause a drastic change. The deployment of solar, wind, battery energy storage and electric cars will need to increase enormously over the next few decades. A massive increase in the production of critical minerals will be required to support such a deployment[705,706].

Mining and processing of minerals is extremely resource intensive. Examples of resources are energy, land, water, chemicals and labor. History has shown that resource intensive processes when carried out at very large scales have serious unintended consequences.

Also, resource intensive processes have high associated costs and can lead to supply concerns. Such concerns are expected to increase with an increasing demand for these materials.

The key concerns about critical minerals are summarized below[707,708,709]:

Concern about energy security: The production and processing of critical minerals is concentrated in fewer geographical locations than fossil fuels[710]. For example, just three countries control the global output for lithium, cobalt and rare earth metals. Moreover, certain countries such as China have an alarmingly large share. Chinese companies have also made large investments in countries–such as Australia, Chile, Democratic Republic of Congo and Indonesia–that have large capabilities for producing critical minerals. In other words, the few countries who currently control the supply of critical minerals could control the global energy supply. This is an energy security risk for the rest of the world.

Concern about environmental impact: Examples of environmental impact related to the production and processing of critical minerals are soil erosion, soil contamination, biodiversity loss, contamination of water bodies by chemicals, reduced surface water storage capacity, hazardous waste, and air pollution from fine particles. A large increase in the production and processing of mineral resources could markedly increase the risk of severe environmental impacts.

Concern about costs and supply: Several factors contribute to the cost and supply risk of minerals. Key factors are highlighted. A decrease in the quality of resources in the future is a significant concern. The extraction of metals from inferior quality ores requires more energy and creates more waste. Thus, deteriorating quality of resources impacts both cost and the environment. Mining projects typically require many years to move from discovery to production–which has significant supply related implications. Also, the massive and long-term need for critical minerals can cause periodic disruptions in energy supply, which can lead to large cost fluctuations.

Summary: The main concerns related to the critical minerals required for a low-carbon transition are energy security, environmental impact, supply risk and cost rise. Any of these can cause serious problems. For example, a few countries control the production and processing of some of the critical minerals. This poses an energy security risk for the rest of the world.

Is recycling an easy solution for the large materials intensity of low-carbon technologies?

Recycling involves the collecting and processing of materials that would otherwise be discarded as trash and turning them into new products. Recycling has several benefits such as reducing waste, conserving resources, and avoiding pollution.

But recycling has some major challenges. Several requirements must be satisfied simultaneously for successful recycling. The requirements include a) an efficient process for collecting and separating materials at the end-of-life, b) a recycling process that can provide the desired material quality with low processing losses, and c) a stable long-term supply of the recycled material.

It is very difficult to satisfy such requirements because of the logistical, technical, and cost challenges. The degree of the challenge depends on the type of material that is being recycled and nature of the application.

The difficulty in achieving high recycle rates is evident from historical data. For reference, we will consider plastic waste and electronic and electrical waste (e-waste) recycling.

According to recent OECD reports, less than 10% of global plastic waste is recycled[711,712]. For reference, the recycling rate for plastics in 2018 was only about 9% in the United States[713].

Recycling of e-waste has also been low. The world produces about 50 tons of e-waste annually. Although e-waste contains expensive materials, only 20% of the e-waste is formally recycled according to the United Nations Environment Programme[714].

What is the current state of recycling for the low-carbon technologies?

In case of solar power, the solar panels require the largest amounts of raw materials. Historically, recycling of solar panels has been very small. For example, only about 10% of the solar panels in the U.S. are recycled[715]. In case of wind power, about 85% of the wind turbine blades can be supposedly recycled[716]. But current recycle rates are very low because of the challenges[717,718].

Moreover, recycling cannot satisfy the massive critical minerals needs of the low-carbon technologies. There are two reasons.

First, the current recycling of critical minerals is far from adequate. For example, less than 1% of lithium is recycled[719].

Second, even if high rates of recycling were miraculously achieved for all critical minerals, only a small fraction of the critical minerals demand can be met by recycling. Why? Because addressing the goals of the Paris Agreement requires a rapid increase in low-carbon technologies and thereby rapid access to large amounts of critical minerals.

Considering the challenges and the historical data from other technologies, the hope that recycling will be an easy solution is not realistic.

Summary: Several requirements must be satisfied simultaneously for successful recycling. The requirements include a) an efficient process for collecting and separating materials at the end-of-life, b) a recycling process that can provide the desired material quality with low processing losses, and c) a stable long-term supply of the recycled material. Examples from history–such as plastics and e-waste–inform that it is extremely difficult to satisfy such requirements because of logistical, technological, and cost challenges.

How does the amount of energy consumed by a country impact its economic well-being?

The gross domestic product per capita–GDP per capita–is an indicator of the economic well-being of the population in a country[720]. GDP per capita for a country is calculated by dividing its GDP by its population during a given time period[721].

A noteworthy relationship exists between the amount of energy consumed per capita by the countries and their economic well-being. Energy consumed per capita for a country is determined by dividing its energy consumption by its population during a given time period[722,723].

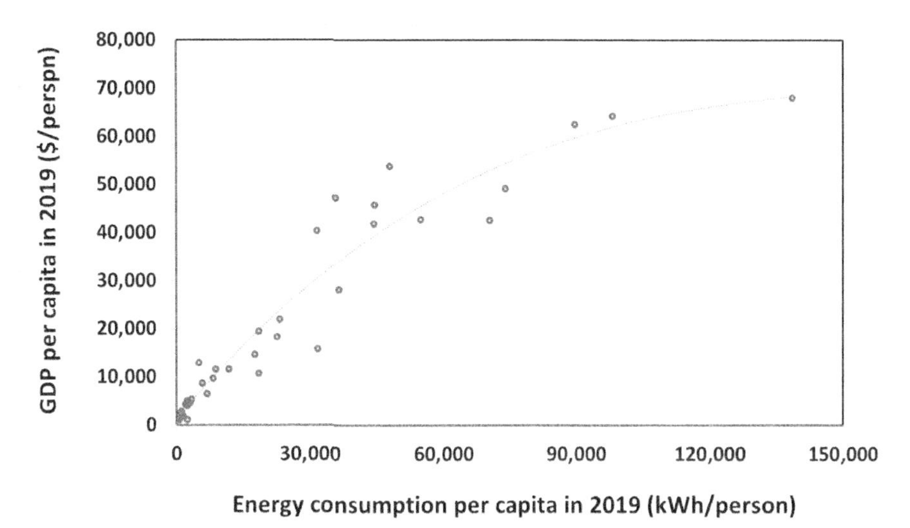

Figure 9.2: GDP per capita vs energy consumption per capita for representative countries in 2019. Source: U.S. EIA and World Bank data.

Figure 9.2 compares representative countries from around the globe for the year 2019. Highlights and implications are discussed below.

- Countries can be divided into three tiers based on their economic well-being: low-income countries with a GDP per capita of less than $20,000 per person, mid-income countries with a GDP per capita between $20,000 and $40,000 per person and high-income countries with a GDP per capita over $40,000 per person.

- Countries with low energy consumed per capita fall in the low-income tier. The general population in these countries is challenged from the viewpoint of its economic well-being. This is expected because high productivity requires large energy consumption. For reference, historical global data shows that a high agricultural output requires access to large amounts of energy[724].
- Majority of the global population resides in low-income countries.
- High-income countries have an energy consumption per capita of over 40,000 kWh/person. For reference, the remaining countries have an average energy consumption per capita of only about 15,000 kWh/year.
- Energy consumed per capita and economic well-being has a linear relationship until a certain energy consumption level, about 50,000 kWh/person, is reached. Energy consumed beyond this level does not significantly improve the country's economic well-being.
- An increase in access to energy will be required to increase the productivity of the low- and mid-income countries. This is crucial for improving their economic well-being.

Summary: Energy consumed by the population of a country is directly correlated to its economic well-being until a certain minimum level of energy consumption is reached. The energy consumed by low-income countries is several-fold lower than this minimum level. Consequently, low-income countries will require a large increase in energy consumption to improve their economic well-being. This has significant implications because majority of the global population resides in low-income countries.

What are the key challenges associated with a low carbon energy transition?

The key challenges can be robustly understood by comparing it to the previous energy transition from traditional biofuels to fossil fuels.

The first key challenge is that the low-carbon transition needs to occur at a high-speed to reduce the impact from climate change[725]. The previous energy transition was not similarly constrained.

The previous transition involved a shift from traditional biofuels to fossil fuels. Around the mid-nineteenth century, fossil fuels contributed to about 20% of the global energy. It took hundred years to increase the global share of fossil fuels to above 80% in the energy mix[726].

Currently, low carbon energy technologies contribute to about 20% of the global energy. A drastic increase in low carbon energy is required to meet the Paris Agreement goals. According to estimates from the International Energy Agency, the share from low carbon energy technologies will need to grow to above 80% over the next thirty years to meet the goals[727]. Thus, the speed of the low-carbon energy transition would need to be three times faster than the previous energy transition.

The second key challenge for the low-carbon energy transition arises from the enormous size of the global energy consumption. Currently, the annual energy consumption is over five times the average annual energy consumption during the previous energy transition[728,729]. Thus, an unprecedented level of deployment of low-carbon technologies will be required over the next decades.

The above two challenges have serious implications for the energy supply and demand balance. Vast resources will be speedily required to accommodate the massive deployment of the low-carbon transition over the next decades.

Unprecedented resource requirements at an ultrafast pace are likely to cause periodic shortages of skilled personnel and critical minerals. The two main reasons for such shortages are competition for qualified personnel and supply chain constraints. Such shortages can disrupt the energy supply. Disruptions in energy supply are also more likely because the low-carbon technologies heavily depend on critical minerals–the supply of which is controlled by a very few countries[730,731]. This has grave implications because even a small disruption in energy supply can cause a large increase in energy price[732].

The third and the most important challenge arises from the lack of a cost and convenience advantage for low-carbon energy compared to the incumbent. The previous energy transition was driven by the distinct cost and convenience advantage offered by fossil fuels over traditional biofuels. However, the low-carbon energy transition is being primarily driven by climate change concerns.

The cost of 24X7 electricity production from renewable technologies is substantially higher than fossil fuel power technologies[733,734,735,736]. Moreover, based on basic scientific principles, renewable technologies will continue to be more costly than fossil fuel technologies for the foreseeable future. Recall, renewable technologies such as solar and wind have a much smaller helping hand from nature compared to fossil fuel technologies and therefore require markedly higher amounts of human activity and resources[737]. This also raises the concern for the environmental impact from renewables when deployed to the same extent as fossil fuels.

The previous energy transition–from traditional biofuels to fossil fuels–was driven by market forces based on the distinct cost and convenience advantage for fossil fuels. There is no cost or convenience advantage for low-carbon energy over fossil fuel technologies. Therefore, the transition from fossil fuels to low carbon energy will need to be driven by government policies–which makes it far more challenging. Why? Because government policies are likely to change substantially with local and global circumstances, for e.g., change in political party, national energy security concerns, natural disasters, and relations between countries.

Summary: The transition to low-carbon energy should be compared with the previous transition from traditional biofuels to fossil fuels to understand the challenges. The low-carbon energy transition has a five times larger scale because of the enormous increase in global energy consumption since the last transition. Also, the low-carbon energy transition needs to be much faster because of the urgency related to climate change. Thus, the low-carbon transition will require unprecedented resources at an ultrafast pace. Furthermore, the low-carbon energy technologies do not have a cost and convenience advantage over the incumbent fossil fuel technologies. This is unlike the prior transition, wherein the superior cost and convenience of fossil fuels over traditional biofuels was the major driver. Consequently, the low-carbon energy transition will need to be driven by government policies from each country, which is extremely challenging.

Why is there so much unrealistic optimism about the low-carbon transition?

Most publications and media focus on the benefits of the low-carbon energy transition and mostly ignore the challenges[738].

The challenges related to the low-carbon energy transition are trivialized by suggesting that they have been mostly addressed or suggesting that they can be easily dealt with further innovations. This has led to unrealistic optimism about the low-carbon transition.

The key reason for the unrealistic optimism is discussed below.

Climate activists have dominated the discussions on low-carbon energy transition. Most climate activists have an extremely strong dislike for fossil fuels. The dislike is mainly a result of their concern about the severe long-term climate impact of fossil fuels.

Why is the extreme dislike for fossil fuels a problem? Because strong negative emotions lead to poor judgement.

The use of the incumbent technology as a reference point is the most reliable approach for understanding the challenges related to a replacement technology. This approach is especially important for complex systems because only a robust reference point can provide an accurate benchmarking. Unfortunately, this crucial approach has been ignored by climate activists because of their extreme dislike for fossil fuels[739].

Over the past several decades, fossil fuels have contributed to over 80% of the global energy consumption[740]. Therefore, fossil fuels must be used as a reference point to understand the challenges related to low-carbon energy solutions. For example, it is crucial to understand how the characteristics of the low-carbon energy transition compare with the characteristics of the previous energy transition to fossil fuels. Recall, the use of fossil fuels as a reference point clearly reveals the severity of the challenges associated with the low-carbon energy transition.

Without the use of fossil fuels as a reference point it is impossible to understand the severity of the challenges associated with the low-carbon energy transition. The extreme dislike towards fossil fuels has led to the disregard for robust methods that would otherwise have allowed for an accurate understanding of the severity of challenges. This, in turn, has led to an unrealistic optimism about the low-carbon energy transition.

A large dose of optimism is necessary for the success of any challenging undertaking. However, optimism is radically different from unrealistic optimism.

Optimism is the belief in the ability to overcome tough challenges. It promotes transparency about the challenges, encourages innovation, collaboration and provides the necessary confidence to address the challenges. These attributes are crucial for a successful outcome.

In case of unrealistic optimism, the challenges are either not acknowledged or their severity is significantly downplayed. Consequently, challenges cannot be successfully addressed.

Why is unrealistic optimism so dangerous? Because it will cause a credibility loss and consequently a backlash from the general population. For example, many climate activists–because of their unrealistic optimism–are claiming that the low-carbon energy transition will have low costs. The true costs will become apparent to the global society as the deployment of low-carbon technologies become more widespread. This will cause a backlash because the true costs will be a shock.

Summary: The unrealistic optimism of climate activists about the low-carbon energy transition is a result of their extreme dislike for fossil fuels. This extreme dislike has led to poor judgment. Specifically, the extreme dislike towards fossil fuels has led to the disregard of robust methods that would otherwise have allowed an accurate understanding of the severity of challenges. Consequently, the severe challenges are either not acknowledged or are significantly downplayed.

Is there consensus amongst global energy experts about the path forward for the energy transition?

Unlike the overwhelming consensus about climate change amongst the global climate experts, there is no consensus amongst global energy experts about an optimal path forward for the low-carbon transition.

Researchers from academia have been the major contributors to the path forward discussions[741]. An overwhelming representation from academia is appropriate for discussing climate change impacts because most climate experts reside in academia[742]. However, it is grossly inadequate for developing path forward proposals about the low-carbon energy transition[743].

Fossil fuels have contributed to about 80% of the global energy since many decades. Correspondingly, most energy experts reside in the fossil fuel industry[744]. The participation of these energy experts has been extremely limited. This is unfortunate because energy experts from the fossil fuel industry have first-hand experience about the challenges related to deployment of massive projects and related practical matters.

Vast personnel and material resources will be required for the low-carbon transition. Energy experts from the fossil industry have a superior understanding about the cost escalations associated with resource-constrained projects, time schedules for complex project deployment, challenges associated with availability of skilled personnel, managing supply chains, and so on[745]. Thus, energy experts from fossil fuel industry will not be unrealistically optimistic about the low-carbon transition[746].

Just like climate experts are essential for understanding climate impacts, energy experts are essential for developing a robust path forward for a low-carbon transition. Consequently, path forward proposals that lack adequate representation of energy experts do not have the required credibility.

Why have the energy industry experts not been included in these discussions? Because of the concern that energy experts with strong financial ties to the fossil fuel industry would find it difficult to be objective[747].

This has led to a severely unfortunate situation–wherein there is a poor representation of energy experts in discussing the future of energy.

Summary: Researchers from academia have been the dominant contributors to the path forward discussions. Expectedly, most academicians have inadequate energy industry experience. On the other hand, energy experts from the fossil fuel industry have substantial experience dealing with the challenges related to deployment of massive projects and related practical matters. This critical experience is transferrable because of the important commonalities within the energy industry. Such experience is crucial for successfully addressing the challenges. Consequently, extensive involvement from these energy experts is critical for developing a robust path forward. Unfortunately, this critical group has been grossly underrepresented in these discussions. Just like climate experts are essential for understanding climate impacts, energy experts are essential for developing a robust path forward for a low-carbon energy transition. Consequently, most of the proposed path forwards lack credibility because of the inadequate expertise of the proposal makers.

...§§§-§-§§§...

10. The Big Picture Discussion

"It is a narrow mind which cannot look at a subject from different points of view."–**George Eliot**

People on the extreme ends have contradictory opinions about the path forward for addressing climate change. Both groups are extremely passionate about the topic. Unfortunately, extreme passion can lead to an emotional state that ignores science and practical matters.

The selection of a judicious path forward is essential for effectively addressing human-caused climate change. How to ensure such a selection? By careful analysis of all crucial aspects and focusing on science and practical matters.

The following approach is used for the big picture discussion. First, the crucial aspects related to climate and energy are summarized. The key requirements for efficiently addressing climate change are discussed next[748]. Finally, a path forward framework which satisfies the key requirements is proposed.

Summary of Crucial Aspects

Climate change impact

The current understanding about human-caused climate change is not perfect because of the extraordinary complexity of the system. Despite the complexity, the global scientific community has made many critical advances[749].

Tens of thousands of scientists have focused their efforts on understanding the climate system. These efforts have been supported by quality data from an array of climate monitoring tools. The collective knowledge from these efforts has showed that human causes are responsible for the significant warming observed over the past several decades.

Many impacts are now understood with high confidence. For example, there is strong evidence for the rising temperatures and sea levels, receding glaciers, ocean acidification, and the increasing frequency and intensity of hot extremes. Clearly, human-caused climate change is a serious problem.

Global climate experts agree that more warming will cause more severe impacts. Also, the experts agree that low-income countries will be impacted more severely because of resource and location constraints.

The consensus is that a fast reduction in greenhouse gases will be required to minimize the global temperature rise. Accordingly, global governments have initiated actions to limit the warning of earth to well below 2°C compared to preindustrial levels.

While the scientific community agrees that the severity of impact will increase with increasing temperature, there is no consensus for 2.5°C or an even 4°C temperature rise being catastrophic for human existence. In other words, the end of the world or doomsday discussions related to climate change have no credibility.

Acceptance of human-caused climate change being a serious problem

Most scientific organizations across the globe have publicly declared that human-caused climate change is a serious problem that must be addressed urgently[750].

Recent polls have revealed that much of the global population accepts that human-caused climate change is a serious problem. But there is a sizable fraction that is not convinced. The misunderstanding has worsened because of inefficient messaging and inappropriate messengers.

Inefficient messaging involves exaggerations about the impact from human-caused climate change which causes overall mistrust about the topic.

Inappropriate messengers include politicians whose polarizing messages promote mistrust. This category also includes celebrities who are perceived to be hypocritical because of their massive carbon footprint.

Options for the power sector

Options that decrease greenhouse gas emissions from the power sector are important because this sector is the largest contributor to greenhouse gases. Also, a low-carbon power sector is critical for efficient electrification.

Many options are available for reducing greenhouse gases from the power sector. However, there is no perfect option. Each option has distinct advantages and challenges (**Table 10.1**)[751].

For example, solar and wind power are low-cost options only at low deployment levels. The intermittency challenge of these technologies becomes more pronounced at high deployment levels. The intermittency challenge can be addressed for wide-scale deployment by including technologies such as energy storage. However, this increases the upfront and total lifetime cost.

The advantages and challenges of the power technologies also depend on the country of deployment. For example, the total lifetime cost for natural gas power is markedly lower in countries that have access to low-cost natural gas.

Option	Key Advantages	Key Challenges
Utility Solar	• Low upfront and total lifetime costs • Low lifecycle greenhouse gas emissions • Low emission of pollutants	• Intermittency challenge. Inability to provide 24X7 electricity. Hence, costs are low **only** at low deployment levels • Land availability is an issue near populated centers
Residential Solar	• Low lifecycle greenhouse gas emissions • Low emission of pollutants • No additional land requirements	• Midrange upfront cost • Midrange-to-high total lifetime cost • Intermittency challenge
Onshore Wind	• Low upfront and total lifetime costs • Low lifecycle greenhouse gas emissions • Low emission of pollutants	• Intermittency challenge. Hence, costs are low **only** at low deployment levels • Land availability can be an issue near populated centers
Offshore Wind	• Low greenhouse gas emissions • Low emission of pollutants • No additional land requirements	• Midrange to high upfront cost • Midrange-to-high total lifetime cost • Intermittency challenge
Solar or wind power with adequate energy storage	• 24X7 electricity generation • Low lifecycle greenhouse gas emissions • Low emission of pollutants	• High upfront cost • High total lifetime cost • Land availability can be an issue near populated centers

Hydropower	• Low-to-midrange upfront and total lifetime costs for suitable locations • Can provide 24X7 electricity • Low greenhouse gas emissions • Low emission of pollutants	• Suitable locations are limited. Project complexity and costs are much higher for sub-optimal locations • Intermittency can be an issue depending upon technology and seasonal changes • Affected population might need relocation
Nuclear Power	• 24X7 electricity generation • Midrange total lifetime costs • Low greenhouse gas emissions • Low emission of pollutants • Very low costs for extending life of an existing plant	• Midrange-to-high upfront costs • Transport and storage of radioactive materials requires special precautions. This challenge will increase with increasing deployment. • Strong negative perception despite good safety record
Biomass power	• 24X7 electricity generation • Low lifecycle greenhouse gas emissions but higher than many other low-carbon technologies	• Midrange upfront cost • Midrange-to-high total lifetime cost • Large water and land requirements
Natural gas power with carbon capture and storage	• 24X7 electricity generation • Low-to-midrange upfront cost • Low-to-midrange lifecycle greenhouse gas emissions	• Midrange-to-high total lifetime costs • Higher greenhouse gas emissions than other low-carbon technologies. • Higher pollutant level than other low-carbon technologies
Coal power with carbon capture and storage	• 24X7 electricity generation • Medium lifecycle greenhouse gas emissions.	• High upfront cost • Midrange-to-high total lifetime cost • Higher emissions than other low-carbon technologies

Natural gas power	• 24X7 electricity generation • Low upfront cost • Low total lifetime costs in certain countries • 50% reduction of greenhouse gas emissions compared to coal power • Much lower pollutants than coal power	• Midrange-to-high total lifetime costs in certain countries. • Higher greenhouse gas emissions than other low-carbon technologies. But further reductions are possible by adding carbon capture and storage • Higher pollutant level than other low-carbon technologies

Table 10.1: Key advantages and challenges of options for decreasing greenhouse gases from the power sector.

Options for the transportation sector

Most efforts related to reducing greenhouse gas emissions have been in the light duty vehicles category. The advantages and challenges of different options for reducing greenhouse gases from this category are summarized in **Table 10.2.**

Option	Key Advantages	Key Challenges
BEVs	• Zero tailpipe emissions • 50% reduction in lifecycle greenhouse emissions based on average global carbon intensity of electrical grid • Low-to-medium total lifetime costs	• High upfront cost • Emission reduction depends on carbon intensity of the electrical grid • Public charging is very slow compared to conventional fueling. • Public charging is expensive compared to residential charging
Advanced biofuels	• 90% reduction in lifecycle greenhouse gas emissions • Refueling does not pose any inconvenience	• High upfront cost • High total lifetime cost • Large land and water requirements
Fuel cell electric vehicles	• Zero tailpipe emissions • Over 40% reduction in lifecycle greenhouse gas emissions	• High upfront cost • High total lifetime cost

	• Refueling rate comparable to conventional gas pumps	• Emission reduction depends on carbon intensity of hydrogen production
Hybrid electric vehicles	• Low-to-midrange upfront cost • Low total lifetime cost • 20-30% reduction in lifecycle greenhouse gas emissions • Excellent compatibility with existing infrastructure	• Modest greenhouse gas emission reductions • Air pollution is only slightly lower than conventional vehicles
Expansion of mass transit	• Low upfront and total lifetime cost because of much lower vehicle miles compared to other options • Over 60% reduction in lifecycle greenhouse gas emissions. • Only option that can drastically decrease road congestion and energy consumption	• Requires reasonably high occupancy for maximizing benefits, 70% occupancy or higher. • Less convenient compared to personal vehicles

Table 10.2: Key advantages and challenges of options for decreasing greenhouse gases from light duty vehicles.

Despite significant progress, BEVs have major challenges. These challenges include high upfront costs and inconvenience. Moreover, the greenhouse gas emissions are sensitive to the carbon intensity of the electrical grid. For example, BEVs reduce greenhouse gases only to a small extent in countries which mainly use coal for electricity generation.

Options such as fuel cell electric vehicles and advanced biofuels suffer from high upfront and lifetime costs. Also, currently they are not established technologies.

Hybrid electric vehicles are superior to battery and fuel cell electric vehicles from a cost and convenience viewpoint. However, emission reductions are modest for this option. Expansion of mass transit option is the most interesting option because of its many critical advantages. But it has a convenience disadvantage compared to using personal vehicles.

What about the other categories in the transportation sector? The overall severity of challenges is much higher in the road freight, shipping, and aviation categories. Consequently, the to-date deployment of low-carbon technologies has been negligible in these categories.

Learnings from fossil fuel technologies and implications

The most severe impact from fossil fuel technologies–climate impact–was definitively understood only after an enormous level of deployment. This has significant implications because the popular low-carbon technologies are currently deployed at extremely low levels.

Nature has transformed the solar energy trapped in plants and other organisms to fossil fuels over millions of years using the heat and pressure in earth. Essentially, nature has eliminated the extreme diluteness and intermittency challenge of solar energy by converting it to high density fossil fuels that are available 24X7 for energy production. Thus, nature has provided a large helping hand for fossil fuel technologies. Nature does not provide such a helping hand to any of the popular low-carbon technologies. Therefore, these technologies require much larger amounts of human activity and resources compared to fossil fuels technologies for supplying equivalent energy.

Global energy needs are gigantic. Consequently, the popular low-carbon technologies when deployed at the required scale will require enormous human activity and unprecedented amounts of resources such as land, minerals, materials, water, etc. Historical data has demonstrated that high human activity and resource requirements cause a high environmental impact.

The above learnings from fossil fuel technologies inform that the popular low-carbon technologies will have some form of severe environmental impact, which will be understood after their wide-scale deployment.

Also, the substantially smaller helping hand from nature for the popular low-carbon technologies implies that their costs will be higher than fossil fuel technologies. Specifically, this learning informs that the cost of solar and wind power will not become lower than fossil fuel power for generating 24X7 electricity in the foreseeable future.

Low-carbon energy transition

Fossil fuels currently provide about 80% of the global energy. Thus, a low-carbon energy transition will require a radical change.

The low-carbon energy transition requires the following: a) transition to low-carbon power, b) electrification and c) decarbonization of difficult-to-electrify categories using technologies such as carbon capture or green hydrogen.

Proven technologies are available for low-carbon power and certain electrification applications, but not for difficult-to-electrify categories. However, even after discounting the difficult-to-electrify categories, there are many critical challenges.

These challenges can be best understood by comparing the characteristics of the previous transition to that of the upcoming transition.

The scale of the low-carbon energy transition is five times larger because of the enormous increase in global energy consumption since the last transition. Also, the low-carbon energy transition needs to be markedly faster because of the urgency related to climate change.

Thus, the low-carbon transition requires an unprecedented scale of technology deployment at an unprecedented speed. This translates to extraordinarily large costs and resources. According to a conservative estimate from a major global consulting firm, an average additional investment of 3.5 trillion dollars per year will be required for thirty years.

The requirement of vast resources at ultrafast speed is likely to cause periodic disruptions in energy supply because of shortage of skilled personnel and materials. Also, the possibility for disruptions in energy supply is high because low-carbon technologies depend on critical minerals whose supply is controlled by a few countries. This is a major concern because a disruption in energy supply increases the energy cost.

Perhaps, the greatest challenge is that the low-carbon technologies do not have a cost and convenience advantage over the incumbent fossil fuel technologies. This is unlike the prior transition, wherein the superior cost and convenience of fossil fuels over traditional biofuels was the major driver. Consequently, the low-carbon energy transition needs to be driven by government policies from each country–which is extremely challenging.

Certain countries have a large share of greenhouse gas emissions. Thirty two countries–China, United States, EU-28, India and Russia–emit about 60% of the global emissions. The rest of the countries emit only 40%. However, this amount is substantial from the viewpoint of achieving net zero global emissions. Thus, concerted efforts from global governments will be required for achieving the target. This is challenging considering the drastically different economic states of countries and geopolitical issues.

The energy consumption per capita of a country is directly correlated to the economic well-being of its population until a certain minimum energy consumption. Majority of the global population resides in low-income countries that consume very little energy. These countries will need to increase energy consumption by several fold to improve their economic well-being. Thus, these low-income countries need abundant low-cost energy.

Climate activists have an extreme dislike towards fossil fuels. This has led to lack of judgment and thereby a disregard for robust methods that can assess the challenges of the low-carbon energy transition. Therefore, climate activists are unaware of the severe challenges and have an unrealistic view of the low-carbon energy transition. This is a major concern because acknowledgment of the challenges is essential for efficiently addressing climate change.

Requirements for efficiently addressing climate change

An energy transition is a massive change. Hence, it will require major policy changes.

These policy changes will impact everyone around the globe. The success of the energy transition will depend on the long-term acceptance of the policies by the general population.

Thus, the primary requirement for efficiently addressing climate change is sustainable support from the global population.

What is required for such a support? Credible policy decisions that are based on science and practical matters.

Such policy decisions will build trust that the global governments are using policies optimally to address climate change. This is crucial because a) an unprecedented spending level will be required for a low-carbon energy transition and b) global governments have a track record of spending tax money in an inefficient manner.

The key requirements for credible policy decisions are discussed below.

- Requirement 1: Policies must achieve the biggest bang for the buck., i.e., target maximum possible benefits at the lowest possible costs[752]. This is critical for a sustainable energy transition.

- Requirement 2: Policies must ensure that there is no disruption in energy supply. This is critical to avoid a large increase in energy price.
- Requirement 3: Policies must be fair across countries and income levels. Policies that place an undue burden will be rejected by the impacted population.
- Requirement 4: Policies must ensure that current and future environmental issues are addressed. For example, policies should ensure that the energy transition does not create severe environmental issues in the future. Failure in this regard will result in a permanent credibility loss.

Path Forward Framework

A robust path forward is crucial for enabling policy decisions that will be acceptable to the global population over a long term. The path forward will be robust only if it satisfies all the key requirements for efficiently addressing climate change. Correspondingly, herein, the discussion is framed around satisfying the key requirements.

The discussion is divided into two parts. In the first part, a generic or global path forward framework is proposed. In the second part, specific examples are discussed to show how the generic path forward framework can enable country-specific policy decisions.

Generic framework

The previous section highlighted the key requirements for efficiently addressing climate change. Here, each requirement is individually addressed to develop a generic path forward framework for an efficient energy transition.

Requirement 1: *Maximum benefits should be achieved at the least possible costs.*

Policies can satisfy this requirement by using a prioritization strategy. A higher priority should be assigned to options that have a high efficiency for greenhouse gas reduction. The deployment of the options should be undertaken in phases based on the prioritization.

Policies should mainly support the deployment of the high priority options in Phase 1, i.e., in the near term. As other options become adequately efficient over the years, they should be given high priority in Phase 2, i.e., after ten years, and so on.

Two prioritization strategies are discussed for identifying high priority options.

The first strategy involves assigning high priority to options that are currently cost-effective in reducing greenhouse gases. Cost-effective options are those that are top quartile performers in terms of upfront and total lifetime costs. This strategy is critical for a fast and impactful reduction of greenhouse gases.

The current cost-effective options for reducing greenhouse gas emissions from the power sector are listed below.

- Utility solar and onshore wind are cost-effective until a certain level of deployment. These technologies are not cost-effective at higher levels because of their intermittency challenge. The cost-effective deployment level depends on factors such as sunshine and wind resource availability, flexibility of other generators in the electrical grid, and connectivity with neighboring electrical grids. For most countries, the cost-effective deployment level for combined solar and wind power is expected to be between 10 and 50%.
- Extending the lifetime of nuclear power plants is very cost effective.
- Hydropower and geothermal power plants are cost-effective when suitable locations are available.
- Biomass power plants are cost-effective when low-cost biomass feedstock is available.
- New generation nuclear power plants are cost-effective in certain countries.
- Natural gas power plants are cost-effective in countries with low-to-moderate natural gas prices[753].

Policies in phase 1 should urgently focus on replacing coal plants with these cost-effective options. This will reduce large quantities of greenhouse gases at a low cost. The cost-effective options are also suitable for meeting the growing electricity demand in developing countries[754].

What about the other options in the power sector? The global governments should provide support for research and development to the other–currently less efficient–options to make them cost competitive. Policies should emphasize these options in the subsequent phases, as they become more cost-efficient.

In the short term, climate enthusiasts should drive the deployment of options that are less cost-effective, such as residential solar. For the purpose of this book, climate enthusiasts are those that believe human-caused climate change is currently the most important problem facing humanity. Most climate enthusiasts are affluent and can afford the extra energy costs.

The cost-effective options for reducing greenhouse gas emissions from the light duty vehicles include expansion of mass transit and hybrid electric vehicles. Expansion of mass transit is very attractive because of its many critical advantages[755]. Correspondingly, policy decisions should urgently focus on a massive expansion of mass transit in Phase 1.

In general, expansion of mass transit must be strongly incentivized over personal vehicles, especially in densely populated regions. In subsequent phases, use of personal vehicles should be increasingly discouraged.

For personal vehicles, hybrid electric vehicles should be the focus of Phase 1. Hybrid electric vehicles have a vastly superior upfront cost compared to BEVs. They also have a lower total lifetime cost in many countries. In the short term, options such as BEVs should be primarily supported by the affluent, for example by climate enthusiasts[756].

An improvement in fuel efficiency is currently the only cost-efficient option for decreasing emissions from road freight, shipping and aviation sectors. Therefore, Phase 1 policies must focus on increasing the fuel efficiency in these sectors. The policies should consider low-carbon technologies for these sectors in the subsequent phases as they become more cost-efficient.

The second strategy involves the deploying of options in a sequence that maximizes the benefits. This strategy ensures that the cart is not put before the horse.

The extent of greenhouse gas reduction via electrification depends on the carbon intensity of the electrical grid. A large benefit requires an electrical grid with low carbon intensity. Recall, the reduction in greenhouse gas emissions related to replacing conventional cars with battery electric cars is less than 25% for a grid with high carbon intensity.

Most of the global population resides in countries where the electrical grid has a mid or high carbon intensity[757]. Consequently, policies in phase 1 should focus on decreasing greenhouse gases from the power sector, i.e., the policy focus should be on decreasing the carbon intensity of electricity generation. Moreover, electrification technologies such as battery electric cars and heat pumps are expensive and/or have a convenience challenge. In the short-term, the affluent climate enthusiasts should voluntarily drive the deployment of electrification technologies such as battery electric cars and heat pumps.

Use of subsidies to promote the deployment of less-efficient solutions–such as residential solar and BEVs–is counterproductive. Such subsidies do not maximize benefits. Moreover, such subsidies are a wasteful gift to the affluent. Why? Because the affluent are primarily using these less-efficient options. Therefore, they are mainly benefiting from the subsides. Clearly, such subsidies are a waste of tax money. Sadly, such ineffective policies are being implemented in many countries.

Requirement 2: *Energy supply should not be disrupted.*
A low-carbon energy transition can be targeted using two distinct approaches.

The first approach involves a destruction of fossil fuels supply by demonizing or stopping oil and gas production. Most climate activists are promoting policies that support this approach.

This approach has a very high risk of disrupting energy supply. Why? Because this approach will cause a premature shortage of oil and gas supply. This approach can only succeed if there is a real-time equivalent increase in the supply of low-carbon energy–which is extremely unlikely based on practical considerations.

The energy supply disruptions resulting from the above approach will cause an extremely negative perception about the low-carbon transition amongst the general population. They will demand a total reversal to fossil fuel energy, which will cause a massive setback to the low-carbon energy transition.

The second approach targets a systematic decrease in the demand for fossil fuels. This can be achieved via policies that provide incentives for efficient low-carbon technologies and lowering the energy consumption. This approach will dramatically reduce the risk of energy supply disruption because it will not cause a premature shortage of oil and gas supply.

Low-carbon energy technologies are materials intensive and have more complex supply chains compared to fossil fuel technologies. Also, the supply of the required critical materials is controlled by a few countries. Thus, critical materials have two major issues: an enormous demand and a high potential for supply constraints. This increases the risk of supply disruption of critical materials and thereby the disruption of energy supply. Energy security is critical for each country. Hence, the deployment speed of low-carbon technologies should be determined by the ability of each country to sustainably acquire critical materials.

An exceedingly large number of personnel with specific skill sets will be required to enable the low-carbon transition. This will result in a shortage of skilled personnel. This can also cause disruptions in energy supply. Consequently, policies should focus on setting up training programs to provide the necessary skill sets. For example, millions of people will need to be trained to avoid shortage of skilled personnel in the United States alone[758].

Requirement 3: *There should be no undue burden on any country or population segment.*

Basic scientific principles dictate that the energy costs will increase with increasing deployment of low-carbon technologies. The magnitude of this cost increase will depend on the energy policies embraced by local and global governments. Low-cost energy is key to improving conditions of the economically challenged. A major fraction of the global population is economically challenged. Policies must ensure that the progress of this category is not negatively impacted by the low-carbon transition. This in turn requires that the cost of the low-carbon transition is minimized, and energy supply is not disrupted. In other words, the policies should satisfy Requirements 1 & 2 to ensure access to low-cost energy for the economically challenged.

Human-caused climate change does not depend on where the CO_2 is released or eliminated. It only depends on the net amount of CO_2 emissions. Therefore, it is ineffective to target high-hanging fruit in United States and EU countries when there is low-hanging fruit in other countries.

In other words, no country should deploy cost-ineffective options. The money can be more effectively used in other countries that can reduce greenhouse gas emissions at a lower cost. This can be achieved by establishing a global fund that maximizes greenhouse gas emission reductions around the globe.

Also, a second global fund must be established to support climate adaptation efforts in developing countries that are expected to be severely impacted by climate change. This is critical because climate mitigation efforts will be too late to help the population in these countries[759].

The contribution share of each country to the global funds should be based on its cumulative greenhouse gas emissions. The contribution level should be such that it does not have a significant economic impact on the country. For example, the maximum annual contribution could be limited to 0.2% of the country's GDP.

Fossil fuel and related industries will suffer enormous job losses. Emphasis should be placed on minimizing the consequences for the impacted population. Accordingly, policies should focus on training the displaced personnel from the affected industries and retooling them with the skill sets required for enabling the low-carbon transition[760].

Requirement 4: *All environmental issues, present and future, should be efficiently addressed.*

Coal produces markedly more greenhouse gases and air pollutants than natural gas. Consequently, policies must focus on urgently replacing coal power plants in Phase 1. Coal power plants should be replaced with cost-effective low-carbon power options or natural gas power plants. Also, policies must focus on decreasing fugitive methane emissions.

The amount of human activity and resources required for an energy technology is directly related to the environmental impact. The popular low-carbon technologies when deployed at the scale necessary to satisfy the global energy needs, will require extraordinarily large amounts of human activity and resources such as land, minerals, materials, water, etc. Thus, the popular low-carbon technologies will have some form of severe environmental impact–which will only be understood after their wide-scale deployment.

The severe environmental impact could be local or global and could involve air, water or land. Currently it is not possible to understand the specific nature of the environmental impact because of the low deployment levels. However, it is likely that the severe impact at wide-scale deployment will limit the further use of the energy technology.

How do we address this problem? Three approaches are discussed below.

Energy policies must focus on maximizing diversity in the energy mix. This is critical for delaying the onset of the severe environmental impact from any low-carbon technology. Maximization of diversity will lead to lower deployment of each technology and thereby easier management of the harmful impacts because of a dilution effect. For reference, if the total share of fossil fuel energy in the past decades had been only 35% instead of 85%, the climate change impact would have been dampened.

Along with solar and wind technologies, other cost-effective low-carbon technologies should also play a significant role in the short and long-term. This is crucial for maximizing the energy diversity.

The limited data on the popular low-carbon technologies has led to an unrealistic optimism about their future environmental impact. Therefore, urgent emphasis should be placed on anticipating the environmental impact of these technologies at the relevant implementation levels. Such information is key to introducing appropriate regulations and developing sustainable resource management solutions in a timely manner, i.e., in the early stages of deployment of the low-carbon technologies. For example, early regulations supporting CO_2 capture and storage would have been helpful.

Policies that move pollution from one region to another should be avoided. Policies that increase net global pollution must be especially avoided. Some countries produce much lower pollutants from oil and gas operations. Policies that restrict oil and gas operations in these countries will increase the net pollution. Why? Because such policies will move the lost oil and gas production to countries with higher emission of pollutants.

Energy policies should discourage energy wastage and encourage options that lower overall energy consumption. For example, policies should strongly incentivize mass transit, i.e., buses, rail and van shuttles. Polices should not promote poor energy habits by incentivizing personal low-carbon vehicles[761]. Also, policies should encourage more efficient use of energy across the whole system.

Country-specific framework

The path forward framework discussed earlier is generically applicable to countries across the globe. However, when applied to an individual country it needs to be fine-tuned based on the country's specific characteristics. Examples of specific characteristics include feedstock price, carbon intensity of electricity generation, population density, and energy consumption per capita[762].

Specific Characteristic: Feedstock Price

United States has a low natural gas price because it has excellent natural gas resources. Consequently, natural gas power plants are the most cost-effective option for producing 24X7 electricity in the United States.

Coal power plants are a major contributor to the greenhouse gas emissions and air pollution in the United States[763]. Thus, the much cleaner natural gas power plants are an attractive option for replacing the coal power plants in Phase 1[764].

An efficient Phase 1 approach for reducing greenhouse gases in the United States would be to replace half of the coal power plants with natural gas power plants and the remainder with cost-effective utility solar power, onshore wind power, hydropower and geothermal power plants[765]. The above approach would cost-effectively reduce greenhouse gases in the United States by 500 million tons per year within a short time frame. In subsequent phases the greenhouse gas emissions could be further reduced by coupling some of the natural gas plants with carbon capture and storage and replacing the remainder with other cost-effective low-carbon options.

Natural gas power plant technology is not an attractive option in countries that have high natural gas prices such as Japan or Sweden or countries such as Belgium and Austria that do not use coal.

Specific Characteristic: Carbon intensity of electrical grid

The electrical grids of countries such as Norway, France and Canada have low carbon intensities–i.e., less than 150 gCO_2 emitted per kWh of electricity generated[766]. Electrification is, therefore, an efficient option for reducing greenhouse gas emissions in such countries. Electrification should be pursued with high priority in these countries assuming the other key requirements are satisfied.

In contrast, the electrical grids of countries such as Germany, United States, India, China and Australia currently have either midrange or high carbon intensities[767]. Policies in these countries should give high priority to reducing greenhouse gases from the power sector in phase 1. Electrification should only be given high priority in subsequent phases after the carbon intensity has been markedly lowered. This is critical to avoid large inefficiencies, i.e., to avoid putting the cart before the horse.

Specific Characteristic: Population density

India, China, EU countries and many others have high population densities. Such countries can substantially benefit from the expansion of mass transit and the corresponding decrease in the personal use of light duty vehicles. The benefits include cost-effective reduction of greenhouse gases, air pollution, road congestion, traffic accidents and energy consumption. Consequently, policies in these countries should strongly incentivize mass transit and discourage the use of personal light duty vehicles.

The above is also true for most urban centers in the United States because of their high population densities. However, there are many regions in the United States which have a low population density. Mass transit alone is not practical in such regions. Policies in the United States should incentivize mass transit in the high-density regions and a combination of mass transit and hybrid electric vehicles in regions with low population density.

Specific Characteristic: Energy consumption per capita

United States has a much higher energy consumption per capita compared to many high-income countries. For example, it has about a two times higher energy consumption per capita than Germany and two and a half times higher than that of the United Kingdom. Consequently, policies in the United States should focus on decreasing the energy consumption per capita by increasing energy efficiency and decreasing energy consumption. Examples include strongly incentivizing mass transit, increased fuel efficiency in the transportation sector, and minimizing energy waste.

On the other end, India has a very low energy consumption per capita. India's energy consumption per capita is factor of ten lower than the United States and factor of three lower than the global average. The low energy consumption is because of the inadequate access to low-cost energy. Improving the well-being of its population will require a large increase in the access to low-cost energy. Addressing the energy shortage in such countries while lowering greenhouse gas emissions will require smart local and global policies. Financial and technology support from advanced economies is critical to incentivize the shift from coal to clean energy in such countries.

Concluding Remarks

There are two predominant views about climate and energy.

One view is that human-caused climate change is not a serious problem. According to this view a low-carbon energy transition is not necessary and fossil fuel use should continue unabated. This path forward is not consistent with the extensive studies by global climate experts. It exposes the global population to an enormous risk from extreme climate change impacts.

The second view is that human-caused climate change is by far the most serious problem facing humankind. According to this view, renewable energy should immediately replace fossil fuels. This path forward is based on an unrealistic optimism about the low-carbon energy transition. Such a path forward will cause major disruptions in energy supply and cost escalations. Consequently, it will be rejected by the global population and cause irreparable damage to climate change mitigation efforts.

Why such opposing views? Because people in the first category are passionate about the advantages of fossil fuels, while the second category are passionate about their dislike for fossil fuels.

There would be more agreement if the passion is redirected towards sustaining the climate and environment while ensuring the well-being of the global population. This would enable productive discussions–which are crucial for meeting the unprecedented challenges posed by the low-carbon transition.

How can the challenges be overcome? By carefully considering climate science, energy fundamentals and practical issues. Specifically, global policies must satisfy certain key requirements to achieve the goal.

- Low-cost solutions with high impact must be given priority[768].
- The deployment of solutions must be sequenced in a manner that maximize the benefits.
- The global energy supply must not be disrupted.
- No undue burden must be placed on any country or population segment.
- Current and future environmental issues must be effectively addressed.

A path forward framework that satisfies these key requirements has been discussed in this book.

Global governments have been inefficient in addressing problems–such as poverty and COVID-19–that require international collaboration. Climate change is a far more challenging problem and will require an unprecedented level of global collaboration.

Robust policy decisions in the early stages, i.e., in Phase 1, are crucial for gaining credibility. The global population will withdraw support for a low-carbon energy transition if the key requirements are not met in Phase 1.

An unrealistic goal–such as a rapid transition–which cannot satisfy the key requirements is very likely to have a poor outcome. It will be far more beneficial to target a goal that can be achieved by policies which satisfy the key requirements[769]. This would ensure sustainable support from the global population and would ensure the best possible outcome.

...§§§-§-§§§...

Glossary & Units

1 trillion = 1000 billion; 1 billion = 1000 million; 1 million = 1000000

1 billion ton = 1 gigaton =1000,000,000 tons

1 MW = 1000 kW; 1 kW = 1000 W

Support from nature: help provided by nature, e.g., tail-wind assistance for air travel.

Capacity factor: The ratio of the electrical energy produced by an electricity generating unit for the period considered to the electrical energy that could have been produced at continuous full power operation during the same period.

Capacity output: The maximum rated output of an electricity generation unit.

Climate resilience: Ability to prepare for, recover from and adapt to detrimental impacts from climate change.

Crude oil: A mixture of hydrocarbons that exists in liquid phase in natural underground reservoirs.

ppm: part per million (10^{-6})

Dispatchable units: Electricity generation units whose output can be varied to follow demand.

Economy of scale: The principle that larger production facilities have lower costs per unit produced than smaller facilities.

Energy density of the source: Amount of energy per appropriate unit (e.g., volume, mass).

Environmental impact: For the purpose of this book, this term includes a collective impact to health, safety and the environment.

Feedstock: Raw material to fuel a machine or industrial process.

Human activity: Actions taken or driven by humans, e.g., clearing of land, production of fertilizers, mining of minerals and so on.

Natural gas: A gaseous mixture of hydrocarbon compounds, the primary one being methane.

Plastics: Polymer (large number of similar chemical units bonded together) compounds with high molecular mass.

Reference technology: The dominant conventional technology

Refinery: An installation that primarily converts crude oil to finished products such as gasoline, heating oil/diesel, propane, butane, etc.

Stratosphere: Second major layer of earth's atmosphere.

Ton: Specifically used in this book as substitute for metric ton (i.e., herein, 1 Ton = 1000 kg)
Travel range: The total distance that can be travelled by the vehicle after it is fully fueled or charged.
Watt (W): Unit of electrical power (e.g., a typical incandescent light bulb: 60-80 W)
Watt hour (Wh): The electrical energy unit of measure equal to one watt of power supplied to, or taken from, an electric circuit steadily for one hour. Example: an average U.S. home uses ~30,000 Wh (30 kWh) per day.

...§§§-§-§§§...

Appendix 1: List of organizations which support that climate change is a serious problem[770].

Academia Chilena de Ciencias, Chile Academia das Ciencias de Lisboa, Portugal Academia de Ciencias de la República Dominicana Academia de Ciencias Físicas, Matemáticas y Naturales de Venezuela Academia de Ciencias Medicas, Fisicas y Naturales de Guatemala Academia Mexicana de Ciencias,Mexico Academia Nacional de Ciencias de Bolivia Academia Nacional de Ciencias del Peru Académie des Sciences et Techniques du Sénégal Académie des Sciences, France Academies of Arts, Humanities and Sciences of Canada Academy of Athens Academy of Science of Mozambique Academy of Science of South Africa Academy of Sciences for the Developing World (TWAS) Academy of Sciences Malaysia Academy of Sciences of Moldova Academy of Sciences of the Czech Republic Academy of Sciences of the Islamic Republic of Iran Academy of Scientific Research and Technology, Egypt Academy of the Royal Society of New Zealand Accademia Nazionale dei Lincei, Italy Africa Centre for Climate and Earth Systems Science African Academy of Sciences Albanian Academy of Sciences Amazon Environmental Research Institute American Academy of Pediatrics American Anthropological Association American Association for the Advancement of Science American Association of State Climatologists (AASC) American Association of Wildlife Veterinarians American Astronomical Society American Chemical Society American College of Preventive Medicine American Fisheries Society	Australian Bureau of Meteorology Australian Coral Reef Society Australian Institute of Marine Science Australian Institute of Physics Australian Marine Sciences Association Australian Medical Association Australian Meteorological and Oceanographic Society Bangladesh Academy of Sciences Botanical Society of America Brazilian Academy of Sciences British Antarctic Survey Bulgarian Academy of Sciences California Academy of Sciences Cameroon Academy of Sciences Canadian Association of Physicists Canadian Foundation for Climate and Atmospheric Sciences Canadian Geophysical Union Canadian Meteorological and Oceanographic Society Canadian Society of Soil Science Canadian Society of Zoologists Caribbean Academy of Sciences views Center for International Forestry Research Chinese Academy of Sciences Colombian Academy of Exact, Physical and Natural Sciences Commonwealth Scientific and Industrial Research Organization (CSIRO) (Australia) Consultative Group on International Agricultural Research Croatian Academy of Arts and Sciences Crop Science Society of America Cuban Academy of Sciences Delegation of the Finnish Academies of Science and Letters Ecological Society of America Ecological Society of Australia Environmental Protection Agency European Academy of Sciences and Arts European Federation of Geologists European Geosciences Union European Physical Society European Science Foundation Federation of American Scientists French Academy of Sciences Geological Society of America

American Geophysical Union American Institute of Biological Sciences American Institute of Physics American Meteorological Society American Physical Society American Public Health Association American Quaternary Association American Society for Microbiology American Society of Agronomy American Society of Civil Engineers American Society of Plant Biologists American Statistical Association Association of Ecosystem Research Centers Australian Academy of Science	Geological Society of Australia Geological Society of London Georgian Academy of Sciences German Academy of Natural Scientists Leopoldina Ghana Academy of Arts and Sciences Indian National Science Academy Indonesian Academy of Sciences Institute of Ecology and Environmental Management
Institute of Marine Engineering, Science and Technology Institute of Professional Engineers New Zealand Institution of Mechanical Engineers, UK InterAcademy Council International Alliance of Research Universities International Arctic Science Committee International Association for Great Lakes Research International Council for Science International Council of Academies of Engineering and Technological Sciences International Research Institute for Climate and Society International Union for Quaternary Research International Union of Geodesy and Geophysics International Union of Pure and Applied Physics Islamic World Academy of Sciences Israel Academy of Sciences and Humanities Kenya National Academy of Sciences Korean Academy of Science and Technology Kosovo Academy of Sciences and Arts l'Académie des Sciences et Techniques du Sénégal Latin American Academy of Sciences Latvian Academy of Sciences Lithuanian Academy of Sciences Madagascar National Academy of Arts, Letters, and Sciences Mauritius Academy of Science and Technology Montenegrin Academy of Sciences and Arts National Academy of Exact, Physical and Natural Sciences, Argentina National Academy of Sciences of Armenia National Academy of Sciences of the Kyrgyz Republic	Palestine Academy for Science and Technology Pew Center on Global Climate Change Polish Academy of Sciences Romanian Academy Royal Academies for Science and the Arts of Belgium Royal Academy of Exact, Physical and Natural Sciences of Spain Royal Astronomical Society, UK Royal Danish Academy of Sciences and Letters Royal Irish Academy Royal Meteorological Society (UK) Royal Netherlands Academy of Arts and Sciences Royal Netherlands Institute for Sea Research Royal Scientific Society of Jordan Royal Society of Canada Royal Society of Chemistry, UK Royal Society of the United Kingdom Royal Swedish Academy of Sciences Russian Academy of Sciences Science and Technology, Australia Science Council of Japan Scientific Committee on Antarctic Research Scientific Committee on Solar-Terrestrial Physics Scripps Institution of Oceanography Serbian Academy of Sciences and Arts Slovak Academy of Sciences Slovenian Academy of Sciences and Arts Society for Ecological Restoration International Society for Industrial and Applied Mathematics Society of American Foresters Society of Biology (UK) Society of Systematic Biologists Soil Science Society of America Sudan Academy of Sciences Sudanese National Academy of Science Tanzania Academy of Sciences

National Academy of Sciences, Sri Lanka National Academy of Sciences, United States of America National Aeronautics and Space Administration National Association of Geoscience Teachers National Association of State Foresters National Center for Atmospheric Research National Council of Engineers Australia National Institute of Water & Atmospheric Research, New Zealand National Oceanic and Atmospheric Administration National Research Council National Science Foundation Natural England Natural Environment Research Council, UK Natural Science Collections Alliance Network of African Science Academies New York Academy of Sciences Nicaraguan Academy of Sciences Nigerian Academy of Sciences Norwegian Academy of Sciences and Letters Oklahoma Climatological Survey Organization of Biological Field Stations Pakistan Academy of Sciences	The Wildlife Society (international) Turkish Academy of Sciences Uganda National Academy of Sciences Union of German Academies of Sciences and Humanities United Nations Intergovernmental Panel on Climate Change University Corporation for Atmospheric Research Woods Hole Oceanographic Institution Woods Hole Research Center World Association of Zoos and Aquariums World Federation of Public Health Associations World Forestry Congress World Health Organization World Meteorological Organization Zambia Academy of Sciences Zimbabwe Academy of Sciences

About the Author

Tushar Choudhary has twenty-five years of R&D experience in addressing environmental issues in the energy field. His experience covers a wide spectrum from fundamental research to technology development & commercialization to technology analysis. Prior to retiring from the energy industry, he served in a dual role as the Director of Technology Analysis & Advancement and Sr. Principal Scientist at a multinational energy company.

He has been granted 40 U.S. & International patents and has published over 50 papers in respected scientific journals. He has given keynote lectures at several International Symposiums and has received numerous prestigious awards within and external to the energy industry. Some of his external awards include the Oklahoma Chemist of the Year Award (American Chemical Society), Global Indus Technovator Award (MIT), and Southwest Industrial Innovation Award (American Chemical Society). He was ranked among the world's top 2% scientists and engineers based on the career impact of his publications (Stanford University database).

Tushar received his Ph.D. in Physical Chemistry from Texas A&M University, College Station; following which he joined the energy industry as a R&D scientist. One of his greatest joys during his time in the energy industry was receiving and sharing knowledge related to all aspects of technology innovation. Following retirement, his goal is to continue sharing knowledge by writing books in areas that he is extremely passionate about.

Also, by this Author: "Critical Comparison of Low Carbon Technologies"

Author Website: https://www.tushar-choudhary.com

Was the book informative? Please leave a review at the retailer website.

References and Notes

[1] IPCC Fourth Assessment Report (2007). Working Group I: The Physical Science Basis. https://archive.ipcc.ch/publications_and_data/ar4/wg1/en/faq-1-2.html

[2] World Meteorological Organization. Our mandate- Observations. https://public.wmo.int/en/our-mandate/what-we-do/observations

[3] NASA Earth Observatory. Record from the deep - Fossil Chemistry. https://earthobservatory.nasa.gov/features/Paleoclimatology_SedimentCores/paleoclimatology_sediment_cores_2.php

[4] U.S. Geological Survey. Paleoclimate Research. https://www.usgs.gov/programs/climate-research-and-development-program/science/paleoclimate-research

[5] NOAA Climate.Gov. How tree rings tell time and climate history. https://www.climate.gov/news-features/blogs/beyond-data/how-tree-rings-tell-time-and-climate-history

[6] National Science Foundation. Ice core Facility. About Ice Cores. https://icecores.org/about-ice-cores

[7] NASA. Global climate change. Core questions- An introduction to ice cores. https://climate.nasa.gov/news/2616/core-questions-an-introduction-to-ice-cores/

[8] IPCC Third Assessment Report (TAR) Climate Change-Synthesis Report (2001). https://www.ipcc.ch/report/ar3/syr/

[9] NOAA Earth system research laboratories. Education files: Natural Climate Change. https://www.esrl.noaa.gov/gmd/education/info_activities/pdfs/TBI_natural_climate_change.pdf

[10] British geological survey. What causes the earth's climate to change? https://www.bgs.ac.uk/discovering-geology/climate-change/what-causes-the-earths-climate-to-change/

[11] J. Eddy. The Maunder minimum. Science, 192, 1189-1202 (1976)

[12] NASA Global Climate change. https://climate.nasa.gov/ask-nasa-climate/2953/there-is-no-impending-mini-ice-age/

[13] NASA Global Climate change. Graphic temperature vs solar activity. https://climate.nasa.gov/climate_resources/189/graphic-temperature-vs-solar-activity/

[14] Milankovitch (Orbital) cycles and their role in earth's climate. https://climate.nasa.gov/news/2948/milankovitch-orbital-cycles-and-their-role-in-earths-climate/

[15] NOAA Climate.gov. Hasn't earth naturally warmed and cooled throughout history? https://www.climate.gov/news-features/climate-qa/hasnt-earth-warmed-and-cooled-naturally-throughout-history

[16] U.S. Geology Survey fact sheet 113-97. The cataclysmic 1991 eruption of Mount Pinatubo, Philippines. https://pubs.usgs.gov/fs/1997/fs113-97/

[17] NASA global climate change. What is the greenhouse effect? https://climate.nasa.gov/faq/19/what-is-the-greenhouse-effect/

[18] NASA Science. Solar system temperatures. https://solarsystem.nasa.gov/resources/681/solar-system-temperatures/

[19] American Chemical Society. Myth: Its water vapor not the CO_2. https://www.acs.org/content/acs/en/climatescience/climatesciencenarratives/its-water-vapor-not-the-co2.html

[20] US EPA: Climate change indicators- Greenhouse gases. https://www.epa.gov/climate-indicators/greenhouse-gases

[21] Note: 25% by oceans and 25% by plants. Reference: https://sos.noaa.gov/catalog/datasets/ocean-atmosphere-co2-exchange/

[22] Greenhouse gas protocol. Global warming potential values. https://www.ghgprotocol.org/sites/default/files/ghgp/Global-Warming-Potential-Values%20%28Feb%2016%202016%29_1.pdf

[23] Note: In this book, "ton" and "metric ton" are used interchangeably.

[24] PBL Netherlands Environmental Assessment Agency 2020 Report. Trends in global CO_2 and GHG emissions. https://www.pbl.nl/en/publications/trends-in-global-co2-and-total-greenhouse-gas-emissions-2020-report

[25] U.S. EPA: Overview of greenhouse gases. https://www.epa.gov/ghgemissions/overview-greenhouse-gases

[26] International Energy Agency (IEA). Methane Tracker 2020. https://www.iea.org/reports/methane-tracker-2020

[27] Greenhouse gas protocol. Global warming potential values. https://www.ghgprotocol.org/sites/default/files/ghgp/Global-Warming-Potential-Values%20%28Feb%2016%202016%29_1.pdf

[28] PBL Netherlands Environmental Assessment Agency 2020 Report. Trends in global CO_2 and GHG emissions. https://www.pbl.nl/en/publications/trends-in-global-co2-and-total-greenhouse-gas-emissions-2020-report

[29] PBL Netherlands Environmental Assessment Agency 2020 Report. Trends in global CO_2 and GHG emissions. https://www.pbl.nl/en/publications/trends-in-global-co2-and-total-greenhouse-gas-emissions-2020-report

[30] PBL Netherlands Environmental Assessment Agency 2020 Report. Trends in global CO_2 and GHG emissions. https://www.pbl.nl/en/publications/trends-in-global-co2-and-total-greenhouse-gas-emissions-2020-report

[31] References cited in the IPCC reports. These reports provide references to the thousands of relevant peer-reviewed scientific publications. https://www.ipcc.ch/reports/

[32] Note: A peer review process involves review of papers by scientific peers. Only papers approved by the scientific peers are published.

[33] Carbon Dioxide Information Analysis Center, Oak Ridge National Laboratory, U.S. Department of Energy, Oak Ridge, Tenn., U.S.A., T.A. Boden, G. Marland, R.J. Andres. 2017. Global, Regional, and National Fossil-Fuel CO_2 Emissions. https://cdiac.ess-dive.lbl.gov/ftp/ndp030/global.1751_2014.ems

[34] IPCC Sixth Assessment Report (AR6). The Physical Science Basis. https://www.ipcc.ch/assessment-report/ar6/

[35] NOAA. Ocean-atmosphere CO_2 exchange. https://sos.noaa.gov/datasets/ocean-atmosphere-co2-exchange/

[36] European Environment Agency. Trends in atmospheric concentrations of CO_2, CH_4 and N_2O. https://www.eea.europa.eu/data-and-maps/daviz/atmospheric-concentration-of-carbon-dioxide-5#tab-chart_5_filters=%7B%22rowFilters%22%3A%7B%7D%3B%22columnFilters%22%3A%7B%22pre_config_polutant%22%3A%5B%22CO2%20(ppm)%22%5D%7D%7D

[37] NOAA Global monitoring laboratory. Trends in atmospheric CO_2. https://gml.noaa.gov/ccgg/trends/

[38] NASA Global climate change. Graphic: the relentless rise of CO_2. Data: Data: Luthi, D., et al. 2008; Etheridge, D.M., et al. 2010; Vostok ice core data/J.R. Petit et al.; NOAA Mauna Loa CO_2 record. https://climate.nasa.gov/climate_resources/24/graphic-the-relentless-rise-of-carbon-dioxide/

[39] NASA Global climate change. Global Temperature. https://climate.nasa.gov/vital-signs/global-temperature/

[40] NASA Global climate change. Global Temperature. https://climate.nasa.gov/vital-signs/global-temperature/

[41] IPCC Fourth Assessment Report, 2007. Palaeoclimate. https://www.ipcc.ch/site/assets/uploads/2018/02/ar4-wg1-chapter6-1.pdf

[42] NASA Global Climate change. Graphic temperature vs solar activity. https://climate.nasa.gov/climate_resources/189/graphic-temperature-vs-solar-activity/

[43] NOAA Climate.gov. Climate models. https://www.climate.gov/maps-data/primer/climate-models

[44] NASA Global Climate change. https://climate.nasa.gov/evidence/

[45] NOAA: Ocean heat content rises. https://www.ncei.noaa.gov/news/ocean-heat-content-rises

[46] NASA. Carbon dioxide fertilization green earth: Study funds. https://www.nasa.gov/feature/goddard/2016/carbon-dioxide-fertilization-greening-earth

[47] IPCC Fifth Assessment Report (AR-5), 2014: Summary for policymakers. https://www.ipcc.ch/site/assets/uploads/2018/02/AR5_SYR_FINAL_SPM.pdf

[48] IPCC Third Assessment Report (TAR), 2007 – The climate system – An overview. https://www.ipcc.ch/site/assets/uploads/2018/03/TAR-01.pdf

[49] J. Fourier. Annales de Chimie et de Physique, 27, 136 (1824). The link is for a different version of his work https://geosci.uchicago.edu/~rtp1/papers/Fourier1827Trans.pdf

[50] J. Tyndall. Philosophical Magazine Series 4, 22 (1861). https://www.tandfonline.com/doi/abs/10.1080/14786446108643138

[51] Svante Arrhenius. Philosophical Magazine Series 5, 41, 251 (1896). https://www.tandfonline.com/doi/abs/10.1080/14786449608620846?src=recsys

[52] Guy Callendar. Quarterly Journal of the Royal Meteorological Society, 64, 223 (1938). https://www.rmets.org/sites/default/files/qjcallender38.pdf

[53] V. Ramanathan, Science, 190, 50 (1975). https://science.sciencemag.org/content/190/4209/50

[54] W. Wang, Science, 194, 685 (1976). https://science.sciencemag.org/content/194/4266/685/tab-article-info

[55] World Meteorological organization. History of IMO. History of WMO. https://public.wmo.int/en/about-us/who-we-are

[56] NASA. Historic missions. Launch of TIROS 1 – World's first weather satellite. https://www.nasa.gov/feature/goddard/2019/launch-of-tiros-1-worlds-1st-weather-satellite-this-week-in-goddard-history-march-31-april-6

[57] NOAA: Celebrating 60 years of the world's first weather satellite. https://www.nesdis.noaa.gov/news/celebrating-60-years-of-the-worlds-first-weather-satellite Note: The site gives credit to NASA for the photo.

[58] NASA Global climate. Earth Science Missions. https://climate.nasa.gov/nasa_science/missions/?page=0&per_page=40&order=title+asc&search=

[59] NASA Global climate. What is the Sun's role in climate change? https://climate.nasa.gov/blog/2910/what-is-the-suns-role-in-climate-change/

[60] Charles Keeling. Tellus, 12, 200 (1960). https://onlinelibrary.wiley.com/doi/abs/10.1111/j.2153-3490.1960.tb01300.x

[61] IPCC Fourth Assessment Report (AR4), 2007. Historical overview of climate change science. https://www.ipcc.ch/report/ar4/wg1/historical-overview-of-climate-change-science/

[62] M. Milankovitch. Royal Serbian academy special publications, section of mathematics and natural science. 132, 633 (1941). https://www.worldcat.org/title/kanon-der-erdbestrahlung-und-seine-anwendung-auf-das-eiszeitenproblem/oclc/10153781

[63] J. Hays, J. Imbrie, N. Shackleton. Science, 194, 1121 (1976). https://science.sciencemag.org/content/194/4270/1121

[64] IPCC Fourth Assessment Report (AR4), 2007. Historical overview of climate change science. https://www.ipcc.ch/report/ar4/wg1/historical-overview-of-climate-change-science/

[65] J. R. Fleming, Eos, Transactions American Geophysical Union 79, 405 (1998).

https://agupubs.onlinelibrary.wiley.com/doi/abs/10.1029/98EO00310

[66] E. Hawkins, P.D. Jones, Quarterly Journal of the Royal Meteorological Society, 139, 1961 (2013). https://www.researchgate.net/publication/251231528_On_increasing_global_temperatures_75_years_after_Callendar https://centaur.reading.ac.uk/32981/1/hawkins_jones_2013.pdf

[67] Note: The only indicator of a significant concern is the recommendation of specific actions by the scientific community to address the concern.

[68] United States President's Science Advisory Committee (1965), Restoring the quality of our environment. https://www.worldcat.org/title/restoring-the-quality-of-our-environment-report-of-the-environmental-pollution-panel-of-the-presidents-science-advisory-committee/oclc/562799

[69] U.S. National Academy of Science Report: Understanding Climate Change (1975). https://archive.org/details/understandingcli00unit/mode/2up

[70] IPCC Reports. https://www.ipcc.ch/reports/

[71] NASA Global Climate Change: Scientific consensus- Earth's climate is warming. https://climate.nasa.gov/scientific-consensus/

[72] U.S. EPA: Atmospheric lifetime and global warming potential defined. https://19january2017snapshot.epa.gov/climateleadership/atmospheric-lifetime-and-global-warming-potential-defined_.html

[73] IPCC Third Assessment Report (TAR), 2007 – Atmospheric chemistry and greenhouse gases. https://www.ipcc.ch/site/assets/uploads/2018/03/TAR-04.pdf

[74] IPCC: About the IPCC. https://www.ipcc.ch/about/

[75] IPCC: Procedure for the preparation, review, acceptance, adoption, approval and publication of IPCC reports. https://archive.ipcc.ch/pdf/ipcc-principles/ipcc-principles-appendix-a.pdf

[76] IPCC: Press release Nov. 2014. Concluding instalment of the fifth assessment report. https://archive.ipcc.ch/pdf/ar5/prpc_syr/11022014_syr_copenhagen.pdf

[77] Note: Individual components of the report were released in Aug. 2021 and Feb. 2022 and April 2022. https://www.ipcc.ch/reports/

[78] IPCC: Reports. https://www.ipcc.ch/reports/

[79] IPCC: Press release Nov. 2014. Concluding instalment of the fifth assessment report. https://archive.ipcc.ch/pdf/ar5/prpc_syr/11022014_syr_copenhagen.pdf
[80] United Nations Climate Change (UNFCC): What is the Kyoto Protocol? https://unfccc.int/kyoto_protocol
[81] United Nations Climate Change (UNFCC): The Paris Agreement. https://unfccc.int/process-and-meetings/the-paris-agreement/the-paris-agreement
[82] United Nations Climate Change (UNFCC): United States of America – First NDC. https://www4.unfccc.int/sites/ndcstaging/PublishedDocuments/United%20States%20of%20America%20First/U.S.A.%20First%20NDC%20Submission.pdf
[83] United Nations Climate Change (UNFCC): China – First NDC. https://www4.unfccc.int/sites/ndcstaging/PublishedDocuments/China%20First/China%27s%20First%20NDC%20Submission.pdf
[84] PBL Netherlands Environmental Assessment Agency 2020 Report. Trends in global CO_2 and GHG emissions. https://www.pbl.nl/en/publications/trends-in-global-co2-and-total-greenhouse-gas-emissions-2020-report
[85] Paris Agreement, United Nations (2015). https://unfccc.int/files/essential_background/convention/application/pdf/english_paris_agreement.pdf
[86] United Nations Climate Change (UNFCC): Paris Agreement – Status of Ratification. https://unfccc.int/process/the-paris-agreement/status-of-ratification
[87] S. Weart, Proceedings of National Academy of Science (U.S.). 110, 3657 (2013). https://www.ncbi.nlm.nih.gov/pmc/articles/PMC3586608/
[88] IPCC Reports: Five comprehensive IPCC reports that have been published to date. https://www.ipcc.ch/reports/
[89] Google Scholar Search: "Climate science". https://scholar.google.com
[90] Google Scholar Search: "Climate change science". https://scholar.google.com
[91] IPCC Comprehensive Reports: TAR-2001, AR4-2007, AR5-2014. https://www.ipcc.ch/reports/
[92] NASA Global climate change. Scientific consensus: Earth's climate is warming. https://climate.nasa.gov/scientific-consensus/

[93] IPCC Reports: https://www.ipcc.ch/reports/

[94] Note: Individual components of the sixth report (AR6) were released in Aug. 2021 and Feb. 2022 and April 2022. The synthesis report will be published late 2022 or early 2023. https://www.ipcc.ch/reports/

[95] NASA Global climate change. Scientific consensus: Earth's climate is warming. https://climate.nasa.gov/scientific-consensus/

[96] Note: Each country has several scientific institutions–Universities and National Laboratories. For example, U.S. has over 400 Ph.D. (doctorate) granting institutions. https://ncses.nsf.gov/pubs/nsf21308/data-tables

[97] Statement from the Commonwealth Academy. https://rsc-src.ca/sites/default/files/Commonwealth%20Academies%20Consensus%20Statement%20on%20Climate%20Change%20-%2012%20March%202018%20-%202.pdf

[98] Statement from the National Academies. https://sites.nationalacademies.org/sites/climate/index.htm

[99] Statement from the American Association for Advancement of Science. https://whatweknow.aaas.org/get-the-facts/

[100] Statement from the American Meteorological society. https://www.ametsoc.org/index.cfm/ams/about-ams/ams-statements/statements-of-the-ams-in-force/climate-change1/

[101] Statement from the Joint Science Academies, https://www.nationalacademies.org/our-work/joint-science-academies-statements-on-global-issues

[102] Statement from the Geology Society of America. https://www.geosociety.org/gsa/positions/position10.aspx

[103] Statement from the Advancing Earth and Space Science (AGU). https://www.agu.org/Share-and-Advocate/Share/Policymakers/Position-Statements/Position_Climate

[104] Statement from the American Chemical Society. https://www.acs.org/content/acs/en/policy/publicpolicies/sustainability/globalclimatechange.html

[105] List of ~200 supporting scientific organizations. http://www.opr.ca.gov/facts/list-of-scientific-organizations.html

[106] Pew Research Center (2014). Political polarization in the American Public. https://www.pewresearch.org/politics/2014/06/12/political-polarization-in-the-american-public/

[107] Pew Research Center (2020). America is exceptional in its nature of political divide. https://www.pewresearch.org/fact-tank/2020/11/13/america-is-exceptional-in-the-nature-of-its-political-divide/

[108] Note: Several books refuting Al Gore's books on climate change have become a rallying cry for Republicans against climate change. Example Book: Inconvenient facts – The science that Al Gore does not want you to know. https://www.amazon.com/INCONVENIENT-FACTS-science-that-doesnt-ebook/dp/B079MLP7DN/

[109] Note: The Green New Deal co-proposed by Alexandria Cortez has made bipartisan conversations about climate science exceedingly challenging because it combines climate policies with a progressive wish list of economic policies.

[110] Inhofe announces climate hypocrite awards: https://www.inhofe.senate.gov/climate-week

[111] The Heritage Foundation. The great hypocrisy of the green new deal. https://www.heritage.org/energy-economics/commentary/the-great-hypocrisy-the-green-new-deal

[112] Science Magazine (October 2020) https://www.sciencemag.org/news/2020/10/trump-has-shown-little-respect-us-science-so-why-are-some-parts-thriving

[113] L. Dillon et al., American Journal of Public Health. https://www.ncbi.nlm.nih.gov/pmc/articles/PMC5922212/

[114] Ten stars who have a colossal carbon footprint. https://www.nzherald.co.nz/entertainment/ten-stars-whove-got-colossal-carbon-footprints/P3V3GSWECCSXA3DHIAT3DIYTTM/

[115] Sustainability: Science, Practice and Policy. The outsized carbon footprints of the super-rich. https://www.tandfonline.com/doi/full/10.1080/15487733.2021.1949847

[116] Private planes, mansions, superyachts. What gives billionaires like Musk and Abramovich such large carbon footprints. https://theconversation.com/private-planes-mansions-and-superyachts-what-gives-billionaires-like-musk-and-abramovich-such-a-massive-carbon-footprint-152514

[117] Splurges of the filthy rich. https://www.gobankingrates.com/money/wealth/splurges-filthy-rich/

[118] Forbes (Aug. 2019). The real reason they behave hypocritically on climate change is because they want to. https://www.forbes.com/sites/michaelshellenberger/2019/08/20/the-real-reason-they-behave-hypocritically-on-climate-change-is-because-they-want-to/?sh=27558a1f185a

[119] Celebrities admit hypocrisy over huge carbon footprint. https://www.energyvoice.com/other-news/209965/celebrities-admit-hypocrisy-over-huge-carbon-footprint/

[120] Breakthrough's discussion papers: Existential Climate related security risk. https://www.breakthroughonline.org.au/papers Note: This paper was widely distributed. It was challenged by scientists who declared that the scientific credibility was low. https://climatefeedback.org/evaluation/iflscience-story-on-speculative-report-provides-little-scientific-context-james-felton/

[121] USA Today. "The world is going to end in 12 years if we do not address climate change". https://www.usatoday.com/story/news/politics/onpolitics/2019/01/22/ocasio-cortez-climate-change-alarm/2642481002/ This statement is not supported by any credible scientific data.

[122] Note: The statements from scientific organizations state that climate change is a serious problem. But there is no consensus or evidence presented which shows that it is a more serious problem currently than other problems such as poverty. In fact, current deaths and hardship that can be directly attributed to climate change are substantially smaller than the impact from poverty.

[123] Food and Agriculture Organization of the United Nations (2020). http://www.fao.org/3/ca8800en/ca8800en.pdf

[124] World Health Organization: Children- improving survival and well-being (2020). https://www.who.int/en/news-room/fact-sheets/detail/children-reducing-mortality

[125] UN Sustainable development goals. https://www.un.org/sustainabledevelopment/sustainable-development-goals/

[126] U.S. Census Bureau (2020 data). https://www.census.gov/newsroom/press-releases/2021/income-poverty-health-insurance-coverage.html

[127] United Nations: No poverty. Why it matters? https://www.un.org/sustainabledevelopment/wp-content/uploads/2018/09/Goal-1.pdf Note: The cost estimate for eliminating global poverty by economists is much lower than the

cost required for just replacing global light duty vehicles with electric vehicles. The global light vehicle segment is only a small contributor (less than 10%) to total greenhouse gas emissions.

[128] Climate change mitigation requires a complete energy transition. Previous energy transitions have required centuries due to the many challenges.

[129] European Commission. Adaption to Climate change. https://ec.europa.eu/clima/policies/adaptation_en

[130] Note: To ensure the optimal use of the resources, it is crucial to prioritize low-carbon solutions in terms of their cost-efficiency and deploy the solutions in a strategic manner. That message is currently missing. Instead, naïve statements are made about low-carbon solutions and their wide-scale deployment that are not consistent with basic scientific and economic principles. These issues will be addressed in detail in later chapters.

[131] C. Lewis, Biomass for the ages. Biomass, 1, 5 (1981). https://www.sciencedirect.com/science/article/abs/pii/0144456581900111

[132] Our World in Data: Energy. https://ourworldindata.org/energy

[133] J. Herbst. The history of transportation. Twenty First Century Books (2006).

[134] R. Berner, The long-term carbon cycle, fossil fuels and atmospheric composition. Nature, 1, 5 (1981). https://www.nature.com/articles/nature02131

[135] J. Nef, An early energy crisis and its consequences. Scientific American, 237, 140 (1977). https://www.jstor.org/stable/24953925?seq=1

[136] R. Hills. Power from steam: A history of stationary steam engines. https://www.cambridge.org/core/books/power-from-steam/8C4164F225F5616682704468F9A4C33D

[137] S. Pain, Power through the ages. Nature (2017). https://www.nature.com/articles/d41586-017-07506-z

[138] E. Wrigley. Energy and the English Industrial revolution, Philosophical Transactions of the Royal Society-A, (2013). https://royalsocietypublishing.org/doi/pdf/10.1098/rsta.2011.0568

[139] Economic history association. The U.S. coal industry in the nineteenth century. https://eh.net/encyclopedia/the-us-coal-industry-in-the-nineteenth-century-2/

[140] P. O'Connor, C. Cleveland. U.S. Energy transitions: 1780-2010. Scientific American, 7, 7955 (2014). https://www.mdpi.com/1996-1073/7/12/7955/htm

[141] U.S. Energy Information Administration. History of energy consumption in the United States. https://www.eia.gov/todayinenergy/detail.php?id=10

[142] Our World in Data: Energy. https://ourworldindata.org/energy

[143] Energy data was obtained from U.S. EIA: https://www.eia.gov/international/data/world/total-energy/total-energy-consumption

[144] BP Statistical review of World energy 2021. https://www.bp.com/content/dam/bp/business-sites/en/global/corporate/pdfs/energy-economics/statistical-review/bp-stats-review-2021-full-report.pdf

[145] Energy data was obtained from U.S. EIA: https://www.eia.gov/international/data/world/total-energy/total-energy-consumption

[146] Note: Liquid petroleum fuels are produced from crude oil.

[147] Historic UK. The great horse manure crisis of 1984. https://www.historic-uk.com/HistoryUK/HistoryofBritain/Great-Horse-Manure-Crisis-of-1894/

[148] K. Dietsche, D. Kuhlgatz. History of the automobile. Fundamentals of automotive and engine technology (2014). https://link.springer.com/chapter/10.1007/978-3-658-03972-1_1

[149] A. Churella. From steam to diesel: Managerial customs and organization capabilities in the twentieth century American automotive industry. Princeton University press (1998) https://www.jstor.org/stable/j.ctt7t2c6

[150] Energy data was obtained from U.S. EIA: https://www.eia.gov/international/data/world/total-energy/total-energy-consumption

[151] BP Statistical review of World energy 2021. https://www.bp.com/content/dam/bp/business-sites/en/global/corporate/pdfs/energy-economics/statistical-review/bp-stats-review-2021-full-report.pdf

[152] NaturalGas.org. History. http://naturalgas.org/overview/history/

[153] T. Considine, R. Watson, M. Blumsack. The economic impact of the Pennsylvania Marcus shale natural gas play. Report by Department of energy and mineral engineering – Penn state University (2010). https://www.researchgate.net/profile/Timothy-

Considine/publication/228367795_The_economic_impacts_of_the_Pennsylvania_Marcellus_Shale_natural_gas_play_An_update/links/56290ddb08ae22b1702ee912/The-economic-impacts-of-the-Pennsylvania-Marcellus-Shale-natura

[154] U.S. Energy.gov. Fossil energy study guide: Natural gas. https://www.energy.gov/sites/prod/files/2017/05/f34/MS_NatGas_Studyguide.pdf

[155] U.S. Energy Information Administration. Natural gas pipelines. https://www.eia.gov/energyexplained/natural-gas/natural-gas-pipelines.php

[156] Our World in Data: Energy. https://ourworldindata.org/energy

[157] American Petroleum Institute (API) Report. Affordable U.S. Energy and U.S. Natural gas and oil (2017). https://www.api.org/-/media/Files/Oil-and-Natural-Gas/Hydraulic-Fracturing/Economic-Benefits/Affordable-US-Energy-and-US-Natural-Gas-and-Oil.pdf

[158] Note: The United States pipeline network has about three million miles of pipelines that connect natural gas production areas and storage facilities with customers.

[159] U.S. Energy Information Administration. Summary statistics for the United States. https://www.eia.gov/electricity/annual/html/epa_01_02.html

[160] Energy data was obtained from U.S. EIA: https://www.eia.gov/international/data/world/total-energy/total-energy-consumption

[161] BP Statistical review of World energy 2021. https://www.bp.com/content/dam/bp/business-sites/en/global/corporate/pdfs/energy-economics/statistical-review/bp-stats-review-2021-full-report.pdf

[162] Note: Using food as an example, energy is required for irrigation pumps, agriculture machinery such as tractors and dryers, climate-controlled storage, transporting food and for producing fertilizers and pesticides. Associated energy costs are included in the cost of food.

[163] Climate watch: Historical GHG emissions. https://www.climatewatchdata.org/ghg-emissions?breakBy=sector&chartType=percentage&end_year=2018&start_year=1990

[164] Note: In absolute terms, the sector generates 37 billion tons of greenhouse gases each year. Total GHG emissions are in terms of

CO_2 equivalent include methane (~3.5 billion tons) and nitrous oxide emissions.

[165] Note: Upfront cost in the power sector is the capital cost for the project. Upfront cost in the transportation sector is the retail price of the vehicle.

[166] Note: History has demonstrated that severity level of environmental impacts is better understood at wide deployment levels. This is because data availability depends on the level of deployment. More the level of deployment, more is the data availability, and more is the understanding of the true severity. Hence, the environmental impact discussion is limited by the current level of deployment of the technology.

[167] IEA Technology Report. Global energy review: CO_2 emissions in 2021. https://www.iea.org/reports/global-energy-review-co2-emissions-in-2021-2

[168] Climate Watch: Historical GHG emissions. https://www.climatewatchdata.org/ghg-emissions?breakBy=regions&end_year=2018&gases=ch4®ions=WORLD%2CWORLD§ors=electricity-heat&start_year=1990

[169] Note: Electrification can reduce greenhouse gases efficiently only if electricity is generated from low-carbon sources. Thus, a transition to low-carbon electricity is a prerequisite for efficient electrification.

[170] U.S. National Renewable Energy Laboratory 2021 Update: Lifecycle greenhouse gas emissions from electricity generation. September 2021. https://www.nrel.gov/docs/fy21osti/80580.pdf

[171] United Nations Economic Commission for Europe. Lifecycle assessments of electricity generation options. https://unece.org/sites/default/files/2022-01/LCA_final-FD_0.pdf

[172] U.S. National Energy technology Laboratory: Life cycle greenhouse gas emissions. Natural gas and power production. https://www.eia.gov/conference/2015/pdf/presentations/skone.pdf

[173] Annex III: Technology specific cost and performance parameters. Climate Change 2014: Mitigation of Climate Change. Contribution of Working Group III to the Fifth Assessment Report of the Intergovernmental Panel on Climate Change. https://www.ipcc.ch/site/assets/uploads/2018/02/ipcc_wg3_ar5_annex-iii.pdf#page=7

[174] U.S. Energy Information Administration: Levelized cost of new generation resources in the annual energy outlook 2022. https://www.eia.gov/outlooks/aeo/pdf/electricity_generation.pdf

[175] Note: Resource-constrained technologies are also known as non-dispatchable technologies.

[176] ScienceDirect: Solar energy. https://www.sciencedirect.com/topics/engineering/solar-energy

[177] U.S. DOE, Energy Efficiency and Renewable Energy. History of solar. https://www1.eere.energy.gov/solar/pdfs/solar_timeline.pdf

[178] Note: One of the early cost breakthroughs came via Exxon Corporation. The breakthrough dropped the cost by a factor of five. See the previous reference, history of solar, for more information.

[179] BP Statistical review of World energy 2020. https://www.bp.com/content/dam/bp/business-sites/en/global/corporate/pdfs/energy-economics/statistical-review/bp-stats-review-2020-full-report.pdf

[180] IEA Global Energy Review 2021. Renewables. https://www.iea.org/reports/global-energy-review-2021/renewables

[181] BP Statistical review of World energy 2021. https://www.bp.com/content/dam/bp/business-sites/en/global/corporate/pdfs/energy-economics/statistical-review/bp-stats-review-2021-full-report.pdf

[182] Our World in Data. H. Ritchie and M. Roser, Renewable energy (2020). https://ourworldindata.org/renewable-energy

[183] Note: History has demonstrated that severity of environmental impacts is better understood at high deployment levels. This is because data availability depends on the level of deployment. More the level of deployment, more is the data availability, and more is the understanding of the true severity of the technology. Hence, the accuracy of the environmental impact discussion is limited by the current level of deployment.

[184] IPCC report: Renewable energy sources and Climate change mitigation. https://www.ipcc.ch/site/assets/uploads/2018/03/SRREN_Full_Report-1.pdf

[185] IPCC, 2014: *Climate Change 2014: Contribution of Working Group III to the Fifth Assessment Report of the Intergovernmental Panel on Climate Change*. Chapter 7. https://www.ipcc.ch/site/assets/uploads/2018/02/ipcc_wg3_ar5_chapter7.pdf

[186] National Renewable Energy Laboratory. Solar photovoltaics technology basics. https://www.nrel.gov/research/re-photovoltaics.html

[187] Photovoltaics report. Fraunhofer Institute for solar energy systems (2020). https://www.ise.fraunhofer.de/content/dam/ise/de/documents/publications/studies/Photovoltaics-Report.pdf

[188] U.S. Energy Information Administration: Cost and performance characteristics of new generation resources in the annual energy outlook 2022. https://www.eia.gov/outlooks/aeo/assumptions/pdf/table_8.2.pdf

[189] U.S. Energy Information Administration: Levelized cost of new generation resources in the annual energy outlook 2022. https://www.eia.gov/outlooks/aeo/pdf/electricity_generation.pdf

[190] International Energy Agency (2021): World Energy model. https://www.iea.org/reports/world-energy-model/techno-economic-inputs

[191] The Energy & Resources Institute (2019): Exploring electricity supply mix scenarios to 2030. https://www.teriin.org/sites/default/files/2019-02/Exploring%20Electricity%20Supply-Mix%20Scenarios%20to%202030.pdf

[192] International Energy Agency: Projected costs of generating electricity, 2020 Edition. https://www.iea.org/reports/projected-costs-of-generating-electricity-2020

[193] U.S. Energy Information Administration: Cost and performance characteristics of new generation resources in the annual energy outlook 2022. https://www.eia.gov/outlooks/aeo/assumptions/pdf/table_8.2.pdf

[194] U.S. Energy Information Administration: Levelized cost of new generation resources in the annual energy outlook 2022. https://www.eia.gov/outlooks/aeo/pdf/electricity_generation.pdf

[195] U.S. EIA: Levelized cost and levelized avoided cost of new generation resources (2020). https://www.eia.gov/outlooks/aeo/pdf/electricity_generation.pdf

[196] National Renewable Energy Laboratory: U.S. Solar photovoltaic system cost benchmark (Q1 2020). https://www.nrel.gov/docs/fy21osti/77324.pdf

[197] U.S. EIA: Levelized cost and levelized avoided cost of new generation resources (2019).

https://www.eia.gov/outlooks/archive/aeo19/pdf/electricity_generation.pdf

[198] Note: Cost provided for solar + storage (CSP) is for concentrated solar thermal power with only 8 hours of storage. The costs will be much higher for a stand-alone system which will require over 24 hours of storage.

[199] Note: Cost provided for solar + storage (CSP) is for concentrated solar thermal power with only 8 hours of storage. The costs will be much higher for a stand-alone system which will require over 24 hours of storage.

[200] National Renewable Energy Laboratory (2015): Overgeneration from solar energy in California. https://www.nrel.gov/docs/fy16osti/65023.pdf

[201] Note: Energy storage is the most common solution. But there are more solutions. This topic is discussed in more details later.

[202] Oakridge National Laboratory (2017): Environmental quality and U.S. power sector- air quality, water quality, land use and environmental justice. https://www.energy.gov/sites/prod/files/2017/01/f34/Environment%20Baseline%20Vol.%202--Environmental%20Quality%20and%20the%20U.S.%20Power%20Sector--Air%20Quality%2C%20Water%20Quality%2C%20Land%20Use%2C%20and%20Environmental%20Justice.pdf

[203] Note: This comparison for land requirements is specifically for the power plant component because the discussion is about localized land constraints for certain regions (e.g., high population urban cities).

[204] U.S. Energy Information Administration: Cost and performance characteristics of new generation resources in the annual energy outlook 2022. https://www.eia.gov/outlooks/aeo/assumptions/pdf/table_8.2.pdf

[205] National Renewable Energy Laboratory: U.S. Solar photovoltaic system cost benchmark (Q1 2020). https://www.nrel.gov/docs/fy21osti/77324.pdf

[206] International Energy Agency: Projected costs of generating electricity, 2020 Edition. https://www.iea.org/reports/projected-costs-of-generating-electricity-2020

[207] The Energy & Resources Institute (2019): Exploring electricity supply mix scenarios to 2030.

https://www.teriin.org/sites/default/files/2019-02/Exploring%20Electricity%20Supply-Mix%20Scenarios%20to%202030.pdf

[208] U.S. DOE, Solar-plus-storage 101. https://www.energy.gov/eere/solar/articles/solar-plus-storage-101

[209] Environmental and Energy Study Institute. Fact sheet: Energy storage (2019). https://www.eesi.org/papers/view/energy-storage-2019#1

[210] National Renewable Energy Laboratory (2012): Renewable electricity futures study. https://www.nrel.gov/docs/fy12osti/52409-2.pdf

[211] U.S. Department of Energy: Energy storage technology and cost characterization report (2019). https://www.energy.gov/sites/prod/files/2019/07/f65/Storage%20Cost%20and%20Performance%20Characterization%20Report_Final.pdf

[212] IEA: Hydropower special market report. https://iea.blob.core.windows.net/assets/4d2d4365-08c6-4171-9ea2-8549fabd1c8d/HydropowerSpecialMarketReport_corr.pdf

[213] Renewable and Sustainable Energy Reviews. Vol. 139, pg. 110705 (2021). https://www.sciencedirect.com/science/article/pii/S1364032121000022

[214] Note: History has demonstrated that severity of environmental impacts is better understood at high deployment levels. This is because data availability depends on the level of deployment. More the level of deployment, more is the data availability, and more is the understanding of the true severity of the technology. Hence, the accuracy of the environmental impact discussion is limited by the current level of deployment.

[215] IPCC report: Renewable energy sources and Climate change mitigation. https://www.ipcc.ch/site/assets/uploads/2018/03/SRREN_Full_Report-1.pdf

[216] Note: Environmental impacts are moderate for hydropower.

[217] U.S. EIA: Solar thermal power plants. https://www.eia.gov/energyexplained/solar/solar-thermal-power-plants.php

[218] U.S. Department of Energy: Energy storage technology and cost characterization report (2019).

https://www.energy.gov/sites/prod/files/2019/07/f65/Storage%20Cost%20and%20Performance%20Characterization%20Report_Final.pdf

[219] Note: A battery system requires several components to facilitate the controlled storage and release of electricity.

[220] World Meteorological Department (WMO). Measuring sunlight. https://public.wmo.int/en/measuring-sunlight

[221] Note: Wind speeds are typically high at night and therefore wind power is a good complement to solar power. Unfortunately, most regions have a very poor combined availability of sunlight and wind for several days in a year. Thus, much more than 24 hours of storage will be required for a scenario wherein 100% electricity is provided by wind and solar power. Current deployment of battery storage, on an average, has an energy storage duration of less than 4 hours because of extremely high costs.

[222] Note: Efficient storage and release of electricity is inherently challenging. Challenges include the stringent requirements of very high cycle efficiency, high energy density and long-term robustness.

[223] IEA Report, revised March 2022. The role of critical minerals on clean energy transitions. https://www.iea.org/reports/the-role-of-critical-minerals-in-clean-energy-transitions

[224] International Energy Agency: Projected costs of generating electricity, 2020 Edition. https://www.iea.org/reports/projected-costs-of-generating-electricity-2020 Note: Representative data for solar thermal is provided in this reference. Moreover, typical energy storage duration is only about 8 hours for solar thermal. Thus, costs for 24-hour storage will be substantially higher.

[225] U.S. Energy Information Administration: Levelized cost of new generation resources in the annual energy outlook 2021. https://www.eia.gov/outlooks/aeo/pdf/electricity_generation.pdf Note: costs are only provided for a hybrid solar + 4-hour battery system. However, the data is a good indicator for the high costs that would be required for a 24-hour battery storage system.

[226] T. V. Choudhary. Critical Comparison of Low Carbon Technologies (October 2020). This reference book provides an estimate for average prices based on representative vehicles. https://www.amazon.com/dp/B08LP8TRLP

[227] U.S. EIA: Levelized cost and levelized avoided cost of new generation resources (2019).

https://www.eia.gov/outlooks/archive/aeo19/pdf/electricity_generation.pdf Note: Representative data for solar thermal is provided in this reference. Note, typical energy storage duration is only about 8 hours for solar thermal. Thus, costs for 24-hour storage will be substantially higher.

[228] U.S. Department of Energy: How do wind turbines work? https://www.energy.gov/eere/wind/how-do-wind-turbines-work

[229] PNAS. Global potential for wind generated electricity. https://www.pnas.org/doi/10.1073/pnas.0904101106

[230] BP Statistical review of World energy 2020. https://www.bp.com/content/dam/bp/business-sites/en/global/corporate/pdfs/energy-economics/statistical-review/bp-stats-review-2020-full-report.pdf

[231] IEA Global Energy Review 2021. Renewables. https://www.iea.org/reports/global-energy-review-2021/renewables

[232] BP Statistical review of World energy 2021. https://www.bp.com/content/dam/bp/business-sites/en/global/corporate/pdfs/energy-economics/statistical-review/bp-stats-review-2021-full-report.pdf

[233] Our World in Data. H. Ritchie and M. Roser, Renewable energy (2020). https://ourworldindata.org/renewable-energy

[234] Note: History has demonstrated that severity of environmental impacts is better understood at high deployment levels. This is because data availability depends on the level of deployment. More the level of deployment, more is the data availability, and more is the understanding of the true severity of the technology. Hence, the accuracy of the environmental impact discussion is limited by the current level of deployment.

[235] IPCC report: Renewable energy sources and Climate change mitigation. https://www.ipcc.ch/site/assets/uploads/2018/03/SRREN_Full_Report-1.pdf

[236] IPCC, 2014: *Climate Change 2014: Contribution of Working Group III to the Fifth Assessment Report of the Intergovernmental Panel on Climate Change*. Chapter 7. https://www.ipcc.ch/site/assets/uploads/2018/02/ipcc_wg3_ar5_chapter7.pdf

[237] U.S. Department of Energy: How a wind turbine works? https://www.energy.gov/articles/how-wind-turbine-works

[238] International Renewable Energy Agency (IRENA): Wind Energy. https://www.irena.org/wind

[239] U.S. Department of Energy: Top ten things you did not know about wind power. https://www.energy.gov/eere/wind/articles/top-10-things-you-didnt-know-about-wind-power

[240] U.S. Energy Information Administration: Cost and performance characteristics of new generation resources in the annual energy outlook 2021. https://www.eia.gov/outlooks/aeo/assumptions/pdf/table_8.2.pdf

[241] U.S. Energy Information Administration: Levelized cost of new generation resources in the annual energy outlook 2021. https://www.eia.gov/outlooks/aeo/pdf/electricity_generation.pdf

[242] International Energy Agency: Projected costs of generating electricity, 2020 Edition. https://www.iea.org/reports/projected-costs-of-generating-electricity-2020

[243] The Energy & Resources Institute (2019): Exploring electricity supply mix scenarios to 2030. https://www.teriin.org/sites/default/files/2019-02/Exploring%20Electricity%20Supply-Mix%20Scenarios%20to%202030.pdf

[244] Note: Capacity factors for resource-constrained technologies vary with region and time. This can significantly complicate the reporting of upfront cost for producing a fixed amount of electricity. To simplify reporting of data, capital costs for electricity generating units are reported per unit of power (for e.g., $/kW). Therefore, the capacity factor for each technology must be considered to estimate the true upfront costs from the reported capital costs.

[245] U.S. Energy Information Administration: Levelized cost of new generation resources in the annual energy outlook 2022. https://www.eia.gov/outlooks/aeo/pdf/electricity_generation.pdf

[246] Note: High end capacity factors are compared to facilitate practical comparisons. The comment from the 2022 U.S. EIA report on total lifetime cost estimation (aka LCOE) related to their assumption of capacity factors is provided for reference purpose: "*EIA evaluated LCOE, LCOS and LACE for each technology based on assumed capacity factors, which generally correspond to the high end of their likely utilization range. This convention is consistent with the use of LCOE and LCOS to evaluate competing technologies in baseload operation such as coal and nuclear plants*"

[247] Note: Even though only about 5% of the total land is directly required for the wind turbines, the entire land area needs to be available when planning for a new wind farm project.

[248] Oakridge National Laboratory (2017): Environmental quality and U.S. power sector- air quality, water quality, land use and environmental justice. https://www.energy.gov/sites/prod/files/2017/01/f34/Environment%20Baseline%20Vol.%202--Environmental%20Quality%20and%20the%20U.S.%20Power%20Sector--Air%20Quality%2C%20Water%20Quality%2C%20Land%20Use%2C%20and%20Environmental%20Justice.pdf

[249] Note: This comparison for land requirements is specifically for the power plant component because the discussion is about localized land constraints for certain regions (e.g., high population urban cities).

[250] U.S. Energy Information Administration: Cost and performance characteristics of new generation resources in the annual energy outlook 2021. https://www.eia.gov/outlooks/aeo/assumptions/pdf/table_8.2.pdf

[251] U.S. Energy Information Administration: Levelized cost of new generation resources in the annual energy outlook 2021. https://www.eia.gov/outlooks/aeo/pdf/electricity_generation.pdf

[252] IEA (2019): World Energy model. https://www.iea.org/reports/world-energy-model/techno-economic-inputs

[253] International Energy Agency: Projected costs of generating electricity, 2020 Edition. https://www.iea.org/reports/projected-costs-of-generating-electricity-2020

[254] U.S. Department of Energy. Offshore wind Research and Development https://www.energy.gov/eere/wind/offshore-wind-research-and-development

[255] U.S. Bureau of Ocean Energy Management: Renewable energy on the outer continental shelf. https://www.boem.gov/renewable-energy/renewable-energy-program-overview

[256] Note: Offshore wind power with energy storage is one of the most expensive solutions due to the high costs related to electricity generation and storage.

[257] Note: Wind power and solar power complement each other and will be used together. This helps, but costs are still high as discussed in the earlier section on solar with energy storage.

[258] U.S. EIA: Hydropower explained. https://www.eia.gov/energyexplained/hydropower/

[259] BP Statistical review of World energy 2021. https://www.bp.com/content/dam/bp/business-sites/en/global/corporate/pdfs/energy-economics/statistical-review/bp-stats-review-2021-full-report.pdf

[260] IPCC report: Renewable energy sources and Climate change mitigation. https://www.ipcc.ch/site/assets/uploads/2018/03/SRREN_Full_Report-1.pdf

[261] Note: History has demonstrated that severity of environmental impacts is better understood at high deployment levels. This is because data availability depends on the level of deployment. More the level of deployment, more is the data availability, and more is the understanding of the true severity of the technology. Hence, the accuracy of the environmental impact discussion is limited by the current level of deployment.

[262] IPCC, 2014: *Climate Change 2014: Contribution of Working Group III to the Fifth Assessment Report of the Intergovernmental Panel on Climate Change*. Chapter 7. https://www.ipcc.ch/site/assets/uploads/2018/02/ipcc_wg3_ar5_chapter7.pdf

[263] U.S. DOE: Waterpower Technologies Office. Types of Hydropower plants. https://www.energy.gov/eere/water/types-hydropower-plants

[264] U.S. Energy Information Administration: Cost and performance characteristics of new generation resources in the annual energy outlook 2021. https://www.eia.gov/outlooks/aeo/assumptions/pdf/table_8.2.pdf

[265] U.S. Energy Information Administration: Levelized cost of new generation resources in the annual energy outlook 2021. https://www.eia.gov/outlooks/aeo/pdf/electricity_generation.pdf

[266] International Energy Agency: Projected costs of generating electricity, 2020 Edition. https://iea.blob.core.windows.net/assets/ae17da3d-e8a5-4163-a3cc-2c6fb0b5677d/Projected-Costs-of-Generating-Electricity-2020.pdf

[267] The Energy & Resources Institute (2019): Exploring electricity supply mix scenarios to 2030. https://www.teriin.org/sites/default/files/2019-02/Exploring%20Electricity%20Supply-Mix%20Scenarios%20to%202030.pdf

[268] IRENA (2019): Renewable generation costs in 2018. International Renewable Energy Agency. https://www.irena.org/-/media/Files/IRENA/Agency/Publication/2019/May/IRENA_Renewable-Power-Generations-Costs-in-2018.pdf

[269] IEA: Hydropower special market report. https://iea.blob.core.windows.net/assets/4d2d4365-08c6-4171-9ea2-8549fabd1c8d/HydropowerSpecialMarketReport_corr.pdf

[270] BP Statistical review of World energy 2021. https://www.bp.com/content/dam/bp/business-sites/en/global/corporate/pdfs/energy-economics/statistical-review/bp-stats-review-2021-full-report.pdf

[271] BP Statistical review of World energy 2021. https://www.bp.com/content/dam/bp/business-sites/en/global/corporate/pdfs/energy-economics/statistical-review/bp-stats-review-2021-full-report.pdf

[272] U.S. EIA: Nuclear power plants. https://www.eia.gov/energyexplained/nuclear/nuclear-power-plants.php

[273] Note: Although nuclear power emits least amount of air pollutants, it is considered to have a low-to-moderate environmental impact due to health and safety concerns related to nuclear radiation. https://www.eia.gov/energyexplained/nuclear/nuclear-power-and-the-environment.php

[274] Note: History has demonstrated that severity of environmental impacts is better understood at high deployment levels. This is because data availability depends on the level of deployment. More the level of deployment, more is the data availability, and more is the understanding of the true severity of the technology. Hence, the accuracy of the environmental impact discussion is limited by the current level of deployment.

[275] IPCC, 2014: *Climate Change 2014: Contribution of Working Group III to the Fifth Assessment Report of the Intergovernmental Panel on Climate Change*. Chapter 7. https://www.ipcc.ch/site/assets/uploads/2018/02/ipcc_wg3_ar5_chapter7.pdf

[276] U.S. EIA: Nuclear explained. https://www.eia.gov/energyexplained/nuclear/

[277] World Nuclear Association: How does a nuclear reactor work? https://www.world-nuclear.org/nuclear-essentials/how-does-a-nuclear-reactor-work.aspx Note: In an enriched form, only about 30 tons of nuclear fuel are required annually. The 300 tons refers to natural Uranium

[278] NEA-OECD Report. Advanced nuclear reactors and future energy markets. https://www.oecd-nea.org/jcms/pl_62463/advanced-nuclear-reactor-systems-and-future-energy-market-needs?details=true

[279] U.S. Energy Information Administration: Cost and performance characteristics of new generation resources in the annual energy outlook 2021. https://www.eia.gov/outlooks/aeo/assumptions/pdf/table_8.2.pdf

[280] U.S. Energy Information Administration: Levelized cost of new generation resources in the annual energy outlook 2021. https://www.eia.gov/outlooks/aeo/pdf/electricity_generation.pdf

[281] International Energy Agency: Projected costs of generating electricity, 2020 Edition. https://www.iea.org/reports/projected-costs-of-generating-electricity-2020

[282] The Energy & Resources Institute (2019): Exploring electricity supply mix scenarios to 2030. https://www.teriin.org/sites/default/files/2019-02/Exploring%20Electricity%20Supply-Mix%20Scenarios%20to%202030.pdf

[283] Note: For example, the capacity factor determines the investment cost for the desired annual electricity generation. Higher capacity factor favors lower cost. Even though nuclear power plants appear to be expensive on a $/kW basis, they are mid-range to high based on the investment required for generating the desired yearly electricity ($/KWh/yr).

[284] U.S. DOE-Office of nuclear energy: Advantages and Challenges of nuclear energy. https://www.energy.gov/ne/articles/advantages-and-challenges-nuclear-energy

[285] Note: This is unfair based on historical data comparing the overall health and safety impact of nuclear power compared to existing technologies. https://ourworldindata.org/nuclear-energy

[286] Note: The incidents are often incorrectly portrayed and give a highly exaggerated view about the dangers associated with nuclear

power. Moreover, nuclear power is often falsely associated with nuclear weapons.

[287] BP Statistical review of World energy 2020. https://www.bp.com/content/dam/bp/business-sites/en/global/corporate/pdfs/energy-economics/statistical-review/bp-stats-review-2020-full-report.pdf

[288] World Bioenergy Association: Global bioenergy statistics, 2019. http://www.worldbioenergy.org/uploads/191129%20WBA%20GBS%202019_HQ.pdf

[289] IPCC report: Renewable energy sources and Climate change mitigation. https://www.ipcc.ch/site/assets/uploads/2018/03/SRREN_Full_Report-1.pdf

[290] Note: History has demonstrated that severity of environmental impacts is better understood at high deployment levels. This is because data availability depends on the level of deployment. More the level of deployment, more is the data availability, and more is the understanding of the true severity of the technology. Hence, the accuracy of the environmental impact discussion is limited by the current level of deployment.

[291] IPCC (2019): Special report on climate change and land. https://www.ipcc.ch/srccl/chapter/summary-for-policymakers/

[292] European Academies Science Advisory Council: Multi-sustainability and functionality in the European Union's forests (2017). https://easac.eu/fileadmin/PDF_s/reports_statements/Forests/EASAC_Forests_web_complete.pdf

[293] National Renewable Energy Laboratory: Biomass energy basics. https://www.nrel.gov/research/re-biomass.html

[294] Note: The life cycle greenhouse gas emissions related to biomass feedstocks are not adequately understood due to the complexity arising from issues such as *land use and change* impact. Thus, the magnitude of impact for greenhouse gas reduction will need to be revisited as the understanding of life cycle greenhouse gas emissions from biomass-to-energy continues to evolve. IPCC (2019): Special report on climate change and land. https://www.ipcc.ch/srccl/chapter/summary-for-policymakers/

[295] U.S. Energy Information Administration: Cost and performance characteristics of new generation resources in the annual energy

outlook 2021.
https://www.eia.gov/outlooks/aeo/assumptions/pdf/table_8.2.pdf

[296] U.S. Energy Information Administration: Levelized cost of new generation resources in the annual energy outlook 2021.
https://www.eia.gov/outlooks/aeo/pdf/electricity_generation.pdf

[297] The Energy & Resources Institute (2019): Exploring electricity supply mix scenarios to 2030.
https://www.teriin.org/sites/default/files/2019-02/Exploring%20Electricity%20Supply-Mix%20Scenarios%20to%202030.pdf

[298] International Energy Agency: Projected costs of generating electricity, 2020 Edition. https://www.iea.org/reports/projected-costs-of-generating-electricity-2020

[299] International Energy Agency Report (Energy Technology Perspectives 2020): Special report on carbon capture utilization and storage. https://iea.blob.core.windows.net/assets/181b48b4-323f-454d-96fb-0bb1889d96a9/CCUS_in_clean_energy_transitions.pdf

[300] General Electric (GE) Company: How a combined cycle plant works. https://www.ge.com/power/resources/knowledge-base/combined-cycle-power-plant-how-it-works

[301] National Energy technology Laboratory: Cost and performance baseline for fossil energy plants (2015).
https://www.netl.doe.gov/energy-analysis/details?id=729

[302] U.S. EIA: Levelized cost and levelized avoided cost of new generation resources (2019).
https://www.eia.gov/outlooks/archive/aeo19/pdf/electricity_generation.pdf

[303] Our World in Data. Natural gas prices.
https://ourworldindata.org/grapher/natural-gas-prices

[304] National Energy technology Laboratory: Cost and performance baseline for fossil energy plants (2015).
https://www.netl.doe.gov/energy-analysis/details?id=729

[305] U.S. EIA: Levelized cost and levelized avoided cost of new generation resources (2019).
https://www.eia.gov/outlooks/archive/aeo19/pdf/electricity_generation.pdf

[306] U.S. EIA, Geothermal explained.
https://www.eia.gov/energyexplained/geothermal/

[307] World Energy Council 2013. Geothermal. https://www.worldenergy.org/assets/images/imported/2013/10/WER_2013_9_Geothermal.pdf

[308] U.S. Department of Energy, History of geothermal energy in America. https://www.energy.gov/eere/geothermal/history-geothermal-energy-america

[309] Geothermal energy throughout the ages. http://www.history.alberta.ca/energyheritage/energy/alternative-energy/geothermal-energy/geothermal-energy-throughout-the-ages.aspx

[310] U.S. EIA, Use of geothermal energy. https://www.eia.gov/energyexplained/geothermal/use-of-geothermal-energy.php

[311] International Energy Agency. Key energy statistics for countries and regions. https://www.iea.org/countries

[312] International Renewable Energy Agency. Geothermal Power. Technology Brief (September 2017). https://www.irena.org/-/media/Files/IRENA/Agency/Publication/2017/Aug/IRENA_Geothermal_Power_2017.pdf

[313] IPCC report: Renewable energy sources and Climate change mitigation. https://www.ipcc.ch/site/assets/uploads/2018/03/SRREN_Full_Report-1.pdf

[314] Note: History has demonstrated that severity of environmental impacts is better understood at high deployment levels. This is because data availability depends on the level of deployment. More the level of deployment, more is the data availability, and more is the understanding of the true severity of the technology. Hence, the accuracy of the environmental impact discussion is limited by the current level of deployment.

[315] U.S. EIA, Use of geothermal energy. https://www.eia.gov/energyexplained/geothermal/use-of-geothermal-energy.php

[316] U.S. DOE, Geothermal. Electricity generation. https://www.energy.gov/eere/geothermal/electricity-generation

[317] Note: The fluid can either be steam or water.

[318] Energy.Gov Site: What is geothermal energy. Please see section 2.4: Technical & non-technical barriers to geothermal development.

https://www.energy.gov/sites/prod/files/2019/05/f63/2-GeoVision-Chap2.pdf

[319] U.S. DOE. Geothermal technologies office. What is an enhanced geothermal system? https://www.energy.gov/sites/default/files/2016/05/f31/EGS%20Fact%20Sheet%20May%202016.pdf

[320] International Renewable Energy Agency. Geothermal power. https://www.irena.org/costs/Power-Generation-Costs/Geothermal-Power

[321] U.S. Energy Information Administration: Cost and performance characteristics of new generation resources in the annual energy outlook 2021. https://www.eia.gov/outlooks/aeo/assumptions/pdf/table_8.2.pdf

[322] U.S. Energy Information Administration: Levelized cost of new generation resources in the annual energy outlook 2021. https://www.eia.gov/outlooks/aeo/pdf/electricity_generation.pdf

[323] U.S. National Academies of Science, Engineering, Medicine. What you need to know about energy: Geothermal. http://needtoknow.nas.edu/energy/energy-sources/renewable-sources/geothermal/

[324] International Energy Agency: Projected costs of generating electricity, 2020 Edition. https://www.iea.org/reports/projected-costs-of-generating-electricity-2020

[325] U.S. DOE. Geothermal technologies office. Geothermal basics. https://www.energy.gov/eere/geothermal/geothermal-basics

[326] U.S. DOE. Geovision 2019. Harnessing the heat below or feet. https://www.energy.gov/sites/default/files/2019/06/f63/GeoVision-full-report-opt.pdf

[327] International Energy Agency. Geothermal power. https://www.iea.org/reports/geothermal-power

[328] U.S. DOE. Geovision 2019. Harnessing the heat below or feet. https://www.energy.gov/sites/default/files/2019/06/f63/GeoVision-full-report-opt.pdf

[329] International Energy Agency. Tracking Transport (2021). https://www.iea.org/reports/tracking-transport-2021

[330] Climate Watch: Historical GHG emissions. https://www.climatewatchdata.org/ghg-emissions?breakBy=regions&end_year=2018&gases=ch4®ions=WORLD%2CWORLD§ors=electricity-heat&start_year=1990

[331] World Resources Institute: Sector by sector. Where do global greenhouse gas emissions come from? https://ourworldindata.org/ghg-emissions-by-sector

[332] International Energy Agency. Transport. Improving the sustainability of passenger and freight transport. https://www.iea.org/topics/transport

[333] U.S. Energy Information Administration. International Energy Outlook 2016. Transportation sector energy consumption. https://www.eia.gov/outlooks/ieo/pdf/transportation.pdf

[334] Note: Mass transit and hybrid electric vehicles will be discussed in a subsequent section because the focus of this chapter is low-cost technology solutions.

[335] Note: Energy efficiency relates to the fraction of the energy in gasoline/diesel or electricity, or hydrogen that can be converted to motion.

[336] Note: Energy density can also be defined as amount of energy available per unit volume.

[337] Third IEEE history of electrotechnology conference. Looking back into electric cars. https://ieeexplore.ieee.org/document/6487583

[338] U.S. Department of Energy. Timeline- History of electric car. https://www.energy.gov/timeline/timeline-history-electric-car

[339] International Energy Agency: Tracking Transport, Electric Vehicles, Tracking Report November 2021. https://www.iea.org/reports/electric-vehicles

[340] Note: The global share of electric car sales was 4.6% in 2020. However, the global share in total global car stock was only about 1%

[341] U.S. Department of Energy. Fueleconomy.com. Where the energy goes: Electric cars. https://www.fueleconomy.gov/feg/atv-ev.shtml

[342] U.S. Department of Energy. Fueleconomy.com. All electric vehicles. https://www.fueleconomy.gov/feg/evtech.shtml

[343] U.S. Energy Information Administration. Few transportation fuels surpass the energy densities of gasoline and diesel. https://www.eia.gov/todayinenergy/detail.php?id=14451

[344] Nissan Website. 2022 Nissan LEAF: Range, charging and battery. https://www.nissanusa.com/vehicles/electric-cars/leaf/features/range-charging-battery.html

[345] Hyundai official website. https://www.hyundaiusa.com/us/en/build

[346] Transport and Environment, Bloomberg NEF. Electric vehicles price parity and phasing out combustion vehicle sales in Europe.

May 2021. Based on this report small and mid-sized electric cars are currently about twice that of conventional cars in Europe (Page 52). https://www.transportenvironment.org/sites/te/files/publications/2021_05_05_Electric_vehicle_price_parity_and_adoption_in_Europe_Final.pdf

[347] T. V. Choudhary. Critical Comparison of Low Carbon Technologies (October 2020). This reference book provides an estimate for average prices based on representative vehicles. https://www.amazon.com/dp/B08LP8TRLP

[348] Note: Prices must be compared for the same trim for an apples-to-apples comparison. Here SEL trim data is presented. Additionally, the Kona Electric (Limited Trim) is about 49% higher than the 2022 Kona conventional (Limited Trim). Prices are directly from the Hyundai retail website. https://www.hyundaiusa.com/us/en/build

[349] Final report for BEUC (The European Consumer Organization)- April 2021: Electric cars: Calculating the total cost of ownership for consumers. https://www.beuc.eu/publications/beuc-x-2021-039_electric_cars_calculating_the_total_cost_of_ownership_for_consumers.pdf

[350] Federal planning Bureau (December 2019): Total cost of ownership of electric cars compared to diesel and gasoline cars in Belgium. https://www.plan.be/uploaded/documents/202001131009220.REP_VHSTCOBEV_12036.pdf

[351] Hsien and Green (MIT): Transition to electric vehicles in China: Implications for total cost of ownership and societal costs. SAE, J. STEEP, 1, 87 (2020). https://doi.org/10.4271/13-01-02-0005

[352] Note: Easy-to-use total cost calculator is available from the U.S. Department of Energy. The site allows the user to choose vehicles for comparison. The calculator shows a moderate higher total cost for BEVs compared to conventional vehicles when compared for the same trim. Example: Hyundai Kona (SEL) vs. Hyundai Kona electric (SEL). https://afdc.energy.gov/calc/

[353] Argonne National Laboratory (April 2021): Comprehensive total cost of ownership calculation for vehicles with different size classes and power trains. Data provided from 2020 to 2050 in Tables B.7b and B.7d https://publications.anl.gov/anlpubs/2021/05/167399.pdf

[354] MIT Energy Initiative, Insights into Future Mobility (November 2019): http://energy.mit.edu/wp-content/uploads/2019/11/Insights-into-Future-Mobility.pdf

[355] U.S. Department of Energy. Electric vehicle benefits and considerations. https://afdc.energy.gov/fuels/electricity_benefits.html

[356] MIT Energy Initiative, Insights into Future Mobility (November 2019): http://energy.mit.edu/wp-content/uploads/2019/11/Insights-into-Future-Mobility.pdf

[357] U.S. DOE. Lifecycle GHG emissions from small sport utility vehicles. https://www.hydrogen.energy.gov/pdfs/21003-life-cycle-ghg-emissions-small-suvs.pdf

[358] IEA: Comparative life cycle GHG emissions from a midsize BEV and ICE vehicle (2021). https://www.iea.org/data-and-statistics/charts/comparative-life-cycle-greenhouse-gas-emissions-of-a-mid-size-bev-and-ice-vehicle

[359] MIT Energy Initiative, Insights into Future Mobility: http://energy.mit.edu/wp-content/uploads/2019/11/Insights-into-Future-Mobility.pdf

[360] U.S. DOE, Office of Energy Efficiency and Renewable Energy. Developing infrastructure to charge plug -in electric vehicles. https://afdc.energy.gov/fuels/electricity_infrastructure.html#dc

[361] Note: Over 80% of charging stations in the United States are Level 2. These charging stations provide only 25 to 70 miles of charging per hour. https://afdc.energy.gov/fuels/electricity_locations.html#/find/nearest?fuel=ELEC

[362] Tesla website: Support section. According to this website, supercharging average session last about 45-50 minutes at city centers. Even if this time decreased to 20 minutes for a full charge, it will still be about a factor of five longer than a conventional gas pump. https://www.tesla.com/support/supercharging

[363] Nissan Website. 2021 Nissan LEAF: Range, charging and battery. https://www.nissanusa.com/vehicles/electric-cars/leaf/features/range-charging-battery.html

[364] Nissan Website. 2021 Nissan LEAF https://www.nissanusa.com/vehicles/electric-cars/leaf.html

[365] Tesla website. Tesla Model 3. https://www.tesla.com/model3/design//overview

[366] Note: Freight vehicles transport goods across state-lines and countries. Therefore, requirements for freight vehicles include the ability to carry large amounts of goods and travel long distances.

[367] U.S. Department of Transportation. Compilation of existing state truck size and weight laws https://ops.fhwa.dot.gov/freight/policy/rpt_congress/truck_sw_laws/app_a.htm#ex17

[368] J.D. Power. How much does a semi-truck weigh? https://www.jdpower.com/cars/shopping-guides/how-much-does-a-semi-truck-weigh

[369] Tesla Website. Semi. https://www.tesla.com/semi

[370] Note: A semi has fuel tank(s) capacity of 200 to 300 gallons of diesel fuel and a fuel economy of about 5 to 7 miles/gallon.

[371] Electric Vehicle Transportation Center: Analysis of fuel cell vehicle developments (2014). http://fsec.ucf.edu/en/publications/pdf/fsec-cr-1987-14.pdf

[372] U.S. Energy Information Administration. Production of Hydrogen. https://www.eia.gov/energyexplained/hydrogen/production-of-hydrogen.php

[373] U.S. Department of Energy. Fuel cells Factsheet. https://www.energy.gov/sites/prod/files/2015/11/f27/fcto_fuel_cells_fact_sheet.pdf

[374] Note: For facilitating this discussion, the box is drawn around the vehicle (tank-to-wheel), i.e., it does not include fuel or electricity production efficiencies.

[375] Toyota Website: 2021 MIRAI Full specs. https://www.toyota.com/mirai/features/mileage_estimates/3002

[376] U.S. Department of Energy. Clean cities- Alternative fuel price report. Currently H_2 price is extremely high because of high delivery costs. Current cost in the United States is about 15$/Kg. 1 Kg of H_2 is roughly equivalent in energy content as 1 gallon of gasoline. https://afdc.energy.gov/files/u/publication/alternative_fuel_price_report_january_2021.pdf

[377] Toyota Website: 2021 MIRAI Full specs. https://www.toyota.com/mirai/features/mileage_estimates/3002

[378] Note: Fuel cell vehicle costs are high and H_2 fuel costs are currently high as well. Per U.S. DOE H_2 price is currently about $15 per gallon of gasoline equivalent. After considering the lower

maintenance costs and higher energy efficiency, the total lifetime costs of fuel cell vehicles are still currently high.

[379] U.S. Department of Energy. Lifecycle greenhouse gas emissions from small sports utility vehicles. https://www.hydrogen.energy.gov/pdfs/21003-life-cycle-ghg-emissions-small-suvs.pdf

[380] U.S. Department of Energy. 5 things to know when fueling up your fuel cell electric vehicle. https://www.energy.gov/eere/articles/5-things-know-when-filling-your-fuel-cell-electric-vehicle

[381] U.S. Energy Information Administration. Few transportation fuels surpass the energy densities of gasoline and diesel. https://www.eia.gov/todayinenergy/detail.php?id=14451

[382] UN Sustainable Development Goal # 2. https://www.un.org/sustainabledevelopment/hunger/

[383] U.S. Department of Energy: Ethanol energy balance. https://afdc.energy.gov/fuels/ethanol_fuel_basics.html#balance

[384] U.S. EPA: Lifecycle Greenhouse Gas Results. https://www.epa.gov/fuels-registration-reporting-and-compliance-help/lifecycle-greenhouse-gas-results ; Note: average emissions data for cellulosic fuels corn stover and switchgrass is provided.

[385] Since the discussion is about wide-scale implementation, waste grease is excluded from advanced biofuels feedstocks. Waste grease has very limited global availability.

[386] IRENA: Advanced Biofuels. What holds them back. November 2019. https://www.irena.org/-/media/Files/IRENA/Agency/Publication/2019/Nov/IRENA_Advanced-biofuels_2019.pdf

[387] U.S. Department of Energy: Biomass Conversion. https://www.energy.gov/sites/prod/files/2016/07/f33/conversion_factsheet.pdf

[388] IEA Bioenergy: Advanced Biofuels- Potential for Cost reduction (2020). https://www.ieabioenergy.com/wp-content/uploads/2020/02/T41_CostReductionBiofuels-11_02_19-final.pdf

[389] European Commission, Joint Research Center (2019). What is limiting the deployment of cellulosic ethanol. https://www.mdpi.com/2076-3417/9/21/4523

[390] European Commission Final Report-Building up the future (2017). https://ec.europa.eu/transparency/regexpert/index.cfm?do=groupDetail.groupDetailDoc&id=33288&no=1

[391] Note: The production scale of an average bioethanol plant (~0.05 million gallons per day) is about 100 times smaller than an averaged sized oil refinery (~5 million gallons per day). This enormous economy-of-scale disadvantage is related to the challenges associated with feedstock supply and process complexity.

[392] IEA Bioenergy: Advanced Biofuels- Potential for Cost reduction (2020). https://www.ieabioenergy.com/wp-content/uploads/2020/02/T41_CostReductionBiofuels-11_02_19-final.pdf

[393] U.S. EIA: Number and capacity of U.S. Refineries. https://www.eia.gov/dnav/pet/pet_pnp_cap1_dcu_nus_a.htm

[394] Cellulosic Ethanol: Status and Innovation. https://www.osti.gov/servlets/purl/1364156

[395] IEA Energy Technology Network. Oil Refineries. https://iea-etsap.org/E-TechDS/PDF/P04_Oil%20Ref_KV_Apr2014_GSOK.pdf

[396] Hydrocarbons-Technology: Saudi Aramco Yanbu Refinery. https://www.hydrocarbons-technology.com/projects/aramco-yanbu/

[397] IEA Bioenergy: Advanced Biofuels- Potential for Cost reduction (2020). https://www.ieabioenergy.com/wp-content/uploads/2020/02/T41_CostReductionBiofuels-11_02_19-final.pdf

[398] European Commission, Joint Research Center (2019). What is limiting the deployment of cellulosic ethanol. https://www.mdpi.com/2076-3417/9/21/4523

[399] European Commission Final Report-Building up the future (2017). https://ec.europa.eu/transparency/regexpert/index.cfm?do=groupDetail.groupDetailDoc&id=33288&no=1

[400] IEA Bioenergy: Advanced Biofuels- Potential for Cost reduction (2020). https://www.ieabioenergy.com/wp-content/uploads/2020/02/T41_CostReductionBiofuels-11_02_19-final.pdf

[401] T. V. Choudhary. Critical Comparison of Low Carbon Technologies (October 2020). This reference book provides an estimate for average prices based on representative vehicles. https://www.amazon.com/dp/B08LP8TRLP

[402] U.S. EPA: Lifecycle Greenhouse Gas Results. https://www.epa.gov/fuels-registration-reporting-and-compliance-

help/lifecycle-greenhouse-gas-results ; Note: average emissions data for cellulosic fuels corn stover and switchgrass is provided.

[403] Hyundai retail website. https://www.hyundaiusa.com/us/en/build

[404] IEA: Comparative life cycle GHG emissions from a midsize BEV and ICE vehicle (2021). https://www.iea.org/data-and-statistics/charts/comparative-life-cycle-greenhouse-gas-emissions-of-a-mid-size-bev-and-ice-vehicle

[405] MIT Energy Initiative, Insights into Future Mobility (November 2019): http://energy.mit.edu/wp-content/uploads/2019/11/Insights-into-Future-Mobility.pdf

[406] Note: For example, the output of dispatchable technologies can be adjusted or energy storage technology can be used to accommodate varied amounts of solar and wind power.

[407] BP statistical review of world energy. https://www.bp.com/en/global/corporate/energy-economics/statistical-review-of-world-energy.html Note: the countries would have eliminated solar and wind, if they had caused undue reliability problems or were unviable.

[408] U.S. Energy Information Administration: Levelized cost of new generation resources in the annual energy outlook 2021. https://www.eia.gov/outlooks/aeo/pdf/electricity_generation.pdf

[409] National Renewable Energy Laboratory (2015): Overgeneration from solar energy in California. https://www.nrel.gov/docs/fy16osti/65023.pdf

[410] National Renewable Energy Laboratory: Ten years of analyzing the duck chart. https://www.nrel.gov/news/program/2018/10-years-duck-curve.html

[411] Note: Due to the crucial need for balancing electricity supply and demand, solar or wind cannot be deployed on a standalone basis. To illustrate this point, consider a small town, which is currently being powered by a combination of coal and natural gas plants. If 100% production of electricity is replaced by solar technology, the town will not receive adequate electricity for a significant fraction of a 24-hour time period (e.g., evening through dawn, or when there is a cloud cover). Therefore, solar and wind need dispatchable technologies such as fossil fuel power plants or nuclear power plants to address their intermittency shortcoming.

[412] Note: A case study is helpful to further discuss the issue. Consider a region, where electricity is provided by natural gas fueled plants, coal fueled plants, and nuclear plants. Due to the dispatchable

nature (ability to match demand with electricity production) of the power plants and the significant flexible features of natural gas power plants, the region can be reliably powered with electricity 24 hours a day and 7 days a week (24X7) for the entire year with the above discussed combination of power plants. Now, let us consider the case where significant amount of solar power is added to replace some of the older coal plants. Solar technology will provide electricity only in the hours when there is sunlight. The other power plants will have to be ramped up or down depending on the electricity demand to address the problems attributed to the non-dispatchable solar plants. In other words, the dispatchable technologies are indirectly subsidizing solar and wind.

[413] OECD and NEA report (2019): System costs with high share of nuclear energy and renewables. https://www.oecd-nea.org/jcms/pl_15000/the-costs-of-decarbonisation-system-costs-with-high-shares-of-nuclear-and-renewables?details=true

[414] OECD and NEA report (2012): The costs of decarbonization. Nuclear Energy and Renewables. System effects in low carbon low carbon electricity systems. https://www.oecd.org/publications/nuclear-energy-and-renewables-9789264188617-en.htm

[415] California ISO: Managing Oversupply. http://www.caiso.com/informed/Pages/ManagingOversupply.aspx

[416] U.S. EIA: California's curtailment of solar electricity generation continues to increase. https://www.eia.gov/todayinenergy/detail.php?id=49276

[417] Note: Estimation based on total 2020 global electricity consumption and 2020 global population data. Electricity generation was converted to electricity consumption using an assumption of 5% loss in transmission and distribution. https://www.bp.com/content/dam/bp/business-sites/en/global/corporate/pdfs/energy-economics/statistical-review/bp-stats-review-2021-full-report.pdf https://data.worldbank.org/indicator/SP.POP.TOTL

[418] California Energy Commission. California solar energy statistics and data. https://ww2.energy.ca.gov/almanac/renewables_data/solar/index_cms.php

[419] U.S. EIA: California's curtailment of solar electricity generation continues to increase. https://www.eia.gov/todayinenergy/detail.php?id=49276

[420] Note: Supplementary technologies such as energy storage or super grids can circumvent this problem. For example, when solar is combined with adequate energy storage, the combination can meet round-the-clock annual electricity demand. Therefore, electricity costs from solar with energy storage can be directly compared with fossil fuel power plants for wide-scale deployment. However, the costs for solar or wind with energy storage are extremely high. For example, the *upfront cost* of solar power plant with adequate energy storage is over **five times** higher than a natural gas power plant.

[421] OECD NEA report: Nuclear Energy and Renewables. System effects in low carbon electricity systems. https://www.oecd.org/publications/nuclear-energy-and-renewables-9789264188617-en.htm

[422] IEA: Projected costs of generating electricity 2020. https://www.iea.org/reports/projected-costs-of-generating-electricity-2020

[423] OECD NEA report (2019): System costs with high share of nuclear energy and renewables. https://www.oecd-nea.org/jcms/pl_15000/the-costs-of-decarbonisation-system-costs-with-high-shares-of-nuclear-and-renewables?details=true

[424] Note: True costs are obtained by including additional costs such as utilization costs, balancing costs and grid costs to the levelized cost of electricity (LCOE). Unfortunately, only LCOE costs are widely communicated in media. Since the profile cost, balancing costs and grid costs are much higher for solar and wind power, the exclusive reporting of LCOE costs by the media is misleading about the true costs of solar and wind power.

[425] OECD NEA report (2019): System costs with high share of nuclear energy and renewables. https://www.oecd-nea.org/jcms/pl_15000/the-costs-of-decarbonisation-system-costs-with-high-shares-of-nuclear-and-renewables?details=true

[426] Hydroelectric Power: A guide for developers and investors. https://www.ifc.org/wps/wcm/connect/906fa13c-2f47-4476-9476-75320e08e5f3/Hydropower_Report.pdf?MOD=AJPERES&CVID=kJQl35z

[427] Water Alternatives, 3, 14 (2010). Lost in Human shadows. The downstream human consequences of dam. https://www.water-alternatives.org/index.php/volume3/v3issue2/80-a3-2-3/file

[428] U.S. Energy.Gov Site: What is geothermal energy. Please see section 2.4: Technical & non-technical barriers to geothermal development. https://www.energy.gov/sites/prod/files/2019/05/f63/2-GeoVision-Chap2.pdf

[429] EMBER: Global electricity review 2022. https://ember-climate.org/insights/research/global-electricity-review-2022/#supporting-material-downloads

[430] Renewables 2021. Global Status Report. https://www.ren21.net/wp-content/uploads/2019/05/GSR2021_Full_Report.pdf

[431] Danish Energy Agency. Integration of wind energy in power systems. A Danish experience (May 2017). https://ens.dk/sites/ens.dk/files/Globalcooperation/integration_of_wind_energy_in_power_systems.pdf

[432] Danish Energy Agency. Integration of wind energy in power systems. A Danish experience. https://ens.dk/sites/ens.dk/files/Globalcooperation/integration_of_wind_energy_in_power_systems.pdf

[433] Ember, 2021 Global Electricity data review. https://ember-climate.org/data/global-electricity/

[434] Energy Facts Norway. Electricity production. https://energifaktanorge.no/en/norsk-energiforsyning/kraftproduksjon/

[435] Danish Energy Agency. Integration of wind energy in power systems. A Danish experience. https://ens.dk/sites/ens.dk/files/Globalcooperation/integration_of_wind_energy_in_power_systems.pdf

[436] Ember, 2021 Global Electricity data review. https://ember-climate.org/data/global-electricity/

[437] U.S. EIA: How much electricity is lost in the transmission and distribution of electricity in the United States? https://www.eia.gov/tools/faqs/faq.php?id=105&t=3

[438] OECD NEA report (2019): System costs with high share of nuclear energy and renewables. https://www.oecd-nea.org/jcms/pl_15000/the-costs-of-decarbonisation-system-costs-with-high-shares-of-nuclear-and-renewables?details=true

[439] Note: Energy storage and/or a overbuild of solar and wind power, and/or extensive transmission grid costs must be included in solar and wind power costs for a valid comparison with fossil fuels. Meaning, when solar and wind power is used in this discussion it includes the various additional components/combinations that are necessary for 24X7 on-demand electricity.

[440] U.S. EIA: Natural gas explained. https://www.eia.gov/energyexplained/natural-gas/

[441] U.S. EIA: Solar explained. https://www.eia.gov/energyexplained/solar/

[442] Note: Intermittency leads to the requirement of energy storage.

[443] International journal of Green Energy, 5,438, 2008. https://www.researchgate.net/publication/233231163_A_Comparison_of_Energy_Densities_of_Prevalent_Energy_Sources_in_Units_of_Joules_Per_Cubic_Meter

[444] Vaclaw Smil. Power densities: A key to understanding energy sources and uses. MIT press, 2015.

[445] Energy policy, 123, 83, 2018. https://www.researchgate.net/publication/327239302_The_spatial_extent_of_renewable_and_non-renewable_power_generation_A_review_and_meta-analysis_of_power_densities_and_their_application_in_the_US

[446] Solar energy cannot be directly transported or stored. Also, a large area is required to harness adequate amounts.

[447] International journal of Green Energy, 5,438, 2008. https://www.researchgate.net/publication/233231163_A_Comparison_of_Energy_Densities_of_Prevalent_Energy_Sources_in_Units_of_Joules_Per_Cubic_Meter

[448] Vaclaw Smil. Power densities: A key to understanding energy sources and uses. MIT press, 2015.

[449] Note: Technology costs do not include cost from externalities such as environmental impact. As discussed later, the environmental impacts from renewable technologies are currently not well understood because of extremely low levels of implementation. Thus, comparison of costs from externalities is not possible currently. The main point in this book is that electricity generation costs in a net zero world will be substantially higher than current costs. This is in contradiction to the messaging from climate activists, who claim that renewable power costs will be lower than fossil fuel power costs.

[450] U.S. Energy Information Administration: Cost and performance characteristics of new generation resources in the annual energy outlook 2021. https://www.eia.gov/outlooks/aeo/assumptions/pdf/table_8.2.pdf

[451] U.S. Energy Information Administration: Levelized cost of new generation resources in the annual energy outlook 2021. https://www.eia.gov/outlooks/aeo/pdf/electricity_generation.pdf

[452] International Energy Agency: Projected costs of generating electricity, 2020 Edition. https://iea.blob.core.windows.net/assets/ae17da3d-e8a5-4163-a3ec-2e6fb0b5677d/Projected-Costs-of-Generating-Electricity-2020.pdf

[453] The Energy & Resources Institute (2019): Exploring electricity supply mix scenarios to 2030. https://www.teriin.org/sites/default/files/2019-02/Exploring%20Electricity%20Supply-Mix%20Scenarios%20to%202030.pdf

[454] T. V. Choudhary. Critical Comparison of Low Carbon Technologies (October 2020). https://www.amazon.com/dp/B08LP8TRLP

[455] Upfront cost of concentrated solar with thermal for only 12 hours of storage is also about five times more than natural gas power plants. This is the low-end cost because storage substantially larger than 12 hours would be required for production of 24x7 on-demand electricity. https://www.iea.org/reports/projected-costs-of-generating-electricity-2020

[456] Note: Energy storage and/or a overbuild of solar and wind power, and/or extensive transmission grid costs must be included in solar and wind power costs for a valid comparison with fossil fuels. Hence, when solar and wind power is used in this discussion it includes the various additional components/combinations that are necessary for 24X7 on-demand electricity.

[457] Note: prior to the technology becoming mature, inefficiencies play a substantial role in determining the cost of the technology. After the technology becomes mature, costs are controlled by basic scientific principles.

[458] Note: Wind and solar could reach maturity within the next few years considering that their deployment levels have increased substantially over the last couple of decades.

[459] Note: Technology costs do not include cost from externalities such as environmental impact. Also, as discussed later, the environmental impacts from renewable technologies are currently not well understood because of extremely low levels of implementation. The main point in this book is that electricity generation costs in a net zero world are going to be substantially higher than current costs.

[460] Note: This only holds true until fossil fuel resources are abundant. Abundant fossil fuel resources are anticipated at-least for the next several decades.

[461] U.S. Department of Energy: Energy storage technology and cost characterization report (2019). https://www.energy.gov/sites/prod/files/2019/07/f65/Storage%20Cost%20and%20Performance%20Characterization%20Report_Final.pdf

[462] IEA: Hydropower special market report. https://iea.blob.core.windows.net/assets/4d2d4365-08c6-4171-9ea2-8549fabd1c8d/HydropowerSpecialMarketReport_corr.pdf

[463] U.S. Department of Energy: Department of energy hydrogen plan. https://www.hydrogen.energy.gov/pdfs/hydrogen-program-plan-2020.pdf

[464] NREL. Energy storage. Days of service sensitivity analysis. https://www.nrel.gov/docs/fy19osti/73520.pdf

[465] U.S. Department of Energy: Energy storage technology and cost characterization report (2019). https://www.energy.gov/sites/prod/files/2019/07/f65/Storage%20Cost%20and%20Performance%20Characterization%20Report_Final.pdf

[466] Note: Current pumped storage hydropower is the most widely used energy storage system globally.

[467] Note: Flow battery technology is more suitable for long duration energy storage compared to Li-ion battery technology. However, it is not a mature technology and therefore has several unknowns.

[468] IEA: The future of hydrogen (2019). https://iea.blob.core.windows.net/assets/9e3a3493-b9a6-4b7d-b499-7ca48e357561/The_Future_of_Hydrogen.pdf

[469] NREL: StoreFAST model overview. Long duration energy storage using H_2 and fuel cells. https://www.energy.gov/sites/default/files/2021-04/h2iq-03242021_2.pdf

[470] IEA Report: Projected costs of generating electricity 2020. https://www.iea.org/reports/projected-costs-of-generating-electricity-2020

[471] International Renewable Energy Agency (IRENA): Energy subsidies. Evolution in the global energy transformation to 2050. https://www.irena.org/publications/2020/Apr/Energy-Subsidies-2020

[472] U.S. EIA. Direct federal financial interventions and subsidies in energy. https://www.eia.gov/analysis/requests/subsidy/pdf/subsidy.pdf

[473] Energy Tracker. Track public money for energy in recovery packages. https://www.energypolicytracker.org

[474] Enerdata. Study on energy subsidies for the European Commission. https://www.enerdata.net/about-us/company-news/european-commission-publishes-enerdata-eu-prices-costs-study.html

[475] Note: Renewable mandates: It requires that a specific share of energy comes from renewables. Net-metering: It is a mechanism that allows the owner of the solar power system to get credit for the electricity they add to the grid. Feed-in-tariffs: a policy that allows the owners of the solar power system to sell electricity at higher-than-normal rates during the contract period.

[476] International Renewable Energy Agency (IRENA): Energy subsidies. Evolution in the global energy transformation to 2050. https://www.irena.org/publications/2020/Apr/Energy-Subsidies-2020

[477] BP Statistical review of World energy 2020. https://www.bp.com/content/dam/bp/business-sites/en/global/corporate/pdfs/energy-economics/statistical-review/bp-stats-review-2020-full-report.pdf Please refer to this report for further information.

[478] U.S. EIA: Cost and performance characteristics of new generating technologies (2020). https://www.eia.gov/outlooks/aeo/assumptions/pdf/table_8.2.pdf

[479] IEA: Role of gas in today's energy transition. https://www.iea.org/reports/the-role-of-gas-in-todays-energy-transitions#key-findings

[480] BP Statistical review of World energy 2021. https://www.bp.com/content/dam/bp/business-sites/en/global/corporate/pdfs/energy-economics/statistical-review/bp-stats-review-2021-full-report.pdf

[481] U.S. National Energy technology Laboratory: Life cycle greenhouse gas emissions. Natural gas and power production. https://www.eia.gov/conference/2015/pdf/presentations/skone.pdf

[482] Annex III: Technology specific cost and performance parameters. Climate Change 2014: Mitigation of Climate Change. Contribution of Working Group III to the Fifth Assessment Report of the Intergovernmental Panel on Climate Change. https://www.ipcc.ch/site/assets/uploads/2018/02/ipcc_wg3_ar5_annex-iii.pdf#page=7

[483] Annex III: Technology specific cost and performance parameters. Climate Change 2014: Mitigation of Climate Change. Contribution of Working Group III to the Fifth Assessment Report of the Intergovernmental Panel on Climate Change. https://www.ipcc.ch/site/assets/uploads/2018/02/ipcc_wg3_ar5_annex-iii.pdf#page=7

[484] IEA: Role of gas in today's energy transition. https://www.iea.org/reports/the-role-of-gas-in-todays-energy-transitions#key-findings

[485] IEA: Natural gas fired power, November 2021. https://www.iea.org/reports/natural-gas-fired-power Note: Although, natural gas is not a long-term solution for electricity generation, nevertheless it would be impractical to not consider it as an intermediate term solution.

[486] U.S. National Renewable Energy Laboratory 2021 Update: Lifecycle greenhouse gas emissions from electricity generation. September 2021. https://www.nrel.gov/docs/fy21osti/80580.pdf

[487] United Nations Economic Commission for Europe (March 2022). Lifecycle assessments of electricity generation options. https://unece.org/sed/documents/2021/10/reports/life-cycle-assessment-electricity-generation-options

[488] IEA: Methane emissions from oil and gas. https://www.iea.org/reports/methane-emissions-from-oil-and-gas

[489] U.S. Energy.gov, Hydraulic fracturing technology. https://www.energy.gov/fecm/hydraulic-fracturing-technology

[490] U.S. Geological Survey: https://www.usgs.gov/mission-areas/water-resources/science/hydraulic-fracturing Note- The actual practice of fracking is only a small part of the overall process of drilling, completing, and producing an oil and gas well.

[491] U.S. EPA. Dec. 16 Report: Hydraulic fracturing for oil and gas: Impacts from the hydraulic fracturing water cycle on drinking

water resources in the United States.
https://cfpub.epa.gov/ncea/hfstudy/recordisplay.cfm?deid=332990

[492] U.S. EPA. 2015 Draft Report: Assessment of the potential impacts of hydraulic fracturing for oil and gas on drinking water resources in the United States.
https://cfpub.epa.gov/ncea/hfstudy/recordisplay.cfm?deid=244651

[493] U.S. Geological Survey: Who we are?
https://www.usgs.gov/about/about-us/who-we-are

[494] U.S. EIA: The distribution of U.S. oil and natural gas wells according to production rates.
https://www.eia.gov/petroleum/wells/pdf/full_report.pdf

[495] U.S. Geological Survey: Hydraulic fracturing.
https://www.usgs.gov/mission-areas/water-resources/science/hydraulic-fracturing?qt-science_center_objects=0#overview

[496] U.S. Geological Survey.
https://www.usgs.gov/search?keywords=hydraulic+fracturing

[497] U.S. Geological Survey, (May 2017).
https://www.usgs.gov/news/national-news-release/unconventional-oil-and-gas-production-not-currently-affecting-drinking

[498] U.S. Geological Survey: Water resources (December 2020).
https://www.usgs.gov/news/bakken-shale-unconventional-oil-and-gas-production-has-not-caused-widespread-hydrocarbon

[499] U.S. Geological Survey: Water resources (July 2019).
https://www.usgs.gov/news/marcellus-shale-natural-gas-production-not-currently-causing-widespread-hydrocarbon

[500] Note: Most chemical and energy-related operations if conducted inappropriately will have substantial environmental consequences. Appropriate regulations can ensure correct operations.

[501] World Health organization (WHO): Air pollution.
https://www.who.int/health-topics/air-pollution#tab=tab_1

[502] Health Effects Institute (HEI): State of global air 2020.
https://www.stateofglobalair.org Note: HEI is funded by U.S. EPA and worldwide motor vehicle industry.

[503] Note: Fine particulate matter (PM2.5), aka fine particles, are responsible for most deaths related to air pollution. Therefore, fine particles are typically used an indicator of air pollution impact. Fine particles are 2.5 microns or less in width. For reference, the average width of human hair is about 75 microns. Fine particles are composed of sulfates, nitrates, ammonia, sodium chloride, black

carbon, mineral dust or water. Over time, the inhalation of these fine particles can cause death because of cardiovascular or respiratory diseases.

[504] Note: For example, there is one academic study which focuses on 2012 data (https://www.sciencedirect.com/science/article/abs/pii/S0013935121000487?via%3Dihub) that claims that the deaths from just one source of air pollution (fossil fuels) are more than the deaths from all sources of air pollution as reported independently by WHO and HEI. Recall, this book focuses on data from governmental and intergovernmental organizations, because of their much higher accountability. A much larger public outcry is expected if the data from WHO is found to be incorrect data as opposed to if data is found to be incorrect from academic research groups. It is unfortunate to see partisan media and books highlight the 2012 study, especially when that is inconsistent with recent studies from credible sources (example: https://www.stateofglobalair.org/data/#/health/plot). Fortunately, UNEP considers the credible recent studies. https://www.unep.org/interactive/air-pollution-note/.

[505] World Health organization (WHO): Ambient (outdoor)Air pollution. https://www.who.int/news-room/fact-sheets/detail/ambient-(outdoor)-air-quality-and-health

[506] Health Effects Institute (HEI): State of global air 2020. https://www.stateofglobalair.org.

[507] Nature Communications, 12,3594, 2021. Source sector and fuel contributions to ambient PM2.5 and attributable mortality to multiple spatial scale. https://www.nature.com/articles/s41467-021-23853-y#Abs1 Note: The study is a collaboration between authors from Washington University, Dalhousie University, Spadaro Environmental Research Consultants, IHME, University of Washington, Pacific Northwest National Laboratory, Department of Atmospheric Sciences, University of Washington, University at Albany, Peking University and University of British Columbia.

[508] Note: As discussed earlier, it is important to consider studies that are consistent with credible studies.

[509] World Health organization (WHO): Tobacco, Key facts. https://www.who.int/news-room/fact-sheets/detail/tobacco

[510] World Health organization (WHO): Obesity. https://www.who.int/news-room/facts-in-pictures/detail/6-facts-on-obesity

[511] World Health organization (WHO): Air pollution. https://www.who.int/health-topics/air-pollution#tab=tab_1

[512] Health Effects Institute (HEI): State of global air 2020. https://www.stateofglobalair.org Note: HEI is funded by U.S. EPA and worldwide motor vehicle industry.

[513] Centre for Disease Control and Prevention (CDC). Road traffic injuries and death: A global problem. https://www.cdc.gov/injury/features/global-road-safety/index.html

[514] Note: This is true for mature technologies that are close to their optimal state.

[515] Note: Conversely, excess investments in energy production cause an oversupply and thereby a decrease in energy costs.

[516] UNDP: Fossil fuel subsidy reforms, lessons and opportunities (2021). https://www.undp.org/publications/fossil-fuel-subsidy-reform-lessons-and-opportunities

[517] IEA: Energy subsidies. https://www.iea.org/topics/energy-subsidies

[518] BP Statistical review of World energy 2021. https://www.bp.com/content/dam/bp/business-sites/en/global/corporate/pdfs/energy-economics/statistical-review/bp-stats-review-2021-full-report.pdf

[519] Note: According to UNDP, most of the subsidies end up with the high-income households, thus defeating the purpose. Reforms to the subsidies are, therefore, important. UNDP: Fossil fuel subsidy reforms, lessons and opportunities (2021). https://www.undp.org/publications/fossil-fuel-subsidy-reform-lessons-and-opportunities

[520] International Monetary Fund. Still not getting energy process right. A global and country update of fossil fuel subsidies. https://www.imf.org/en/Publications/WP/Issues/2021/09/23/Still-Not-Getting-Energy-Prices-Right-A-Global-and-Country-Update-of-Fossil-Fuel-Subsidies-466004

[521] Note: IMF has released several such reports. Typically, there is an update each year. The previous reference is for the latest update.

[522] Note: Subsidies are defined as financial incentives from the government. Therefore, the cost from external factors does not fit the basic definition of subsidies. This is important because headlines from such reports give the false impression that the

governments are providing trillions of dollars of funding for fossil fuels.

[523] U.S. EPA: Benefits and cost of the clean air act 1990-2020. https://www.epa.gov/clean-air-act-overview/benefits-and-costs-clean-air-act-1990-2020-second-prospective-study

[524] International Monetary Fund. Still not getting energy process right. A global and country update of fossil fuel subsidies. https://www.imf.org/en/Publications/WP/Issues/2021/09/23/Still-Not-Getting-Energy-Prices-Right-A-Global-and-Country-Update-of-Fossil-Fuel-Subsidies-466004

[525] Note: At the bare minimum, the report should emphasize that the road congestion and accidents are related to personal transportation and do not depend on the vehicle technology. This should be highlighted several times in the report.

[526] U.S. DOE, Office of Energy Efficiency and Renewable Energy, Electric-drive vehicles, 2017, https://afdc.energy.gov/files/u/publication/electric_vehicles.pdf

[527] Argonne National Laboratory (April 2021): Comprehensive total cost of ownership calculation for vehicles with different size classes and power trains. https://publications.anl.gov/anlpubs/2021/05/167399.pdf

[528] MIT Energy Initiative, Insights into Future Mobility (November 2019): http://energy.mit.edu/wp-content/uploads/2019/11/Insights-into-Future-Mobility.pdf

[529] Note: Easy-to-use total lifetime cost calculator is available from the U.S. Department of Energy. The site allows the user to choose vehicles for comparison. https://afdc.energy.gov/calc/

[530] Toyota official website: https://www.toyota.com/camry/

[531] IEA: Well to wheels greenhouse gas emissions for cars by power trains. (2021). https://www.iea.org/data-and-statistics/charts/well-to-wheels-greenhouse-gas-emissions-for-cars-by-powertrains
Note: Life cycle GHG emissions were estimated by adding vehicle manufacturing emissions to well-to-wheels emissions. A second IEA report provides similar information: Comparative life cycle GHG emissions from a midsize BEV and ICE vehicle (2021). https://www.iea.org/data-and-statistics/charts/comparative-life-cycle-greenhouse-gas-emissions-of-a-mid-size-bev-and-ice-vehicle

[532] Note: Toyota Prius Eco has a 40% superior fuel efficiency than Toyota Corolla. Toyota RAV4 hybrid has a 30% higher efficiency than Toyota RAV4.

[533] MIT Energy Initiative, Insights into Future Mobility (November 2019): http://energy.mit.edu/wp-content/uploads/2019/11/Insights-into-Future-Mobility.pdf

[534] U.S. DOE. Alternative fuels data center. Emissions from hybrid and plug in electric vehicles. https://afdc.energy.gov/vehicles/electric_emissions.html

[535] IEA: Comparative life cycle GHG emissions from a midsize BEV and ICE vehicle (2021). https://www.iea.org/data-and-statistics/charts/comparative-life-cycle-greenhouse-gas-emissions-of-a-mid-size-bev-and-ice-vehicle

[536] Note: For reference the carbon intensity of the United States electrical grid is about 385 gCO_2/kWh. https://www.eia.gov/tools/faqs/faq.php?id=74&t=11

[537] U.S. Department of Transportation. Public transportation's role in responding to climate change (2010). https://www.transit.dot.gov/sites/fta.dot.gov/files/docs/PublicTransportationsRoleInRespondingToClimateChange2010.pdf

[538] Note: Results from the U.S. Department of Transportation provide quality historical data about the impact of vehicle occupancy in buses/trains on the CO_2 emissions. The study showed that for bus transit, the CO_2 emissions decreased from 291 gCO_2 per passenger mile for the 28% (observed) average historical vehicle occupancy of the bus to 82 gCO_2 per passenger mile for 100% (maximum) occupancy of the bus. For commuter train, the CO_2 emissions decreased from 150 gCO_2 per passenger mile for the 30% (observed) average historical occupancy of the commuter train to 45 gCO_2 per passenger mile for 100% (maximum) occupancy. The study also showed an average CO_2 emission of 327 gCO_2/per passenger mile for private auto trips based on the observed average occupancy of 1.4 passengers per private auto trip for work and other general reasons.

[539] IPCC AR6: Climate change 2022. Mitigation of climate change. Summary for policy makers. Figure SPM.7: Overview of mitigation options and their estimated range of costs and potentials in 2030.

[540] T. V. Choudhary. Critical Comparison of Low Carbon Technologies (October 2020). https://www.amazon.com/dp/B08LP8TRLP

[541] Note: When the total travel miles are low, fuel and maintenance costs are also low.

[542] Ford official website. 2022 Ford transit van (base cost: $47,000). https://www.ford.com/trucks/transit-passenger-van-wagon/ A typical shuttle van has a total occupancy for about 15 passengers. This represents a practical size for ensuring high occupancy levels for most trips. In contrast it can be challenging to ensure high occupancy levels for a bus with 40+ passenger capacity.

[543] Note: Substantially fewer miles for mass transit compared to private light duty vehicles equals fewer vehicles, lower fuel consumption and lower maintenance and therefore, lower upfront and total lifetime costs.

[544] Note: Herein, the discussion is focused on light duty BEVs. These are the primary focus of current global subsidies.

[545] Note: Very few batteries electric vehicle options exist for large vehicles currently. Considering that BEVs are about 40 to 50% more expensive than conventional vehicles, large vehicles are too expensive for most people to buy. For example, assuming a 45,000$ average price for a large vehicle, the corresponding BEV will average around 65,000$.

[546] Note: Comparison data is estimated using U.S. Fueleconomy.gov for fuel and electricity cycle emissions and International Energy Agency for vehicle manufacturing emissions. Estimation = Addition of the fuel and electricity cycle emissions from U.S. Fueleconomy.gov to the vehicle manufacturing cycle emissions from IEA data. Most of the contribution (80%) is from fuel and electricity cycle emissions. This data was directly included from the Fueleconomy.gov site provided in the next reference.

[547] Fueleconomy.gov. https://www.fueleconomy.gov/feg/Find.do?action=sbsSelect

[548] IEA: Comparative life cycle GHG emissions from a midsize BEV and ICE vehicle (2021). https://www.iea.org/data-and-statistics/charts/comparative-life-cycle-greenhouse-gas-emissions-of-a-mid-size-bev-and-ice-vehicle

[549] U.S. DOE. Lifecycle GHG emissions from small sport utility vehicles. https://www.hydrogen.energy.gov/pdfs/21003-life-cycle-ghg-emissions-small-suvs.pdf

[550] MIT Energy Initiative, Insights into Future Mobility (November 2019): http://energy.mit.edu/wp-content/uploads/2019/11/Insights-into-Future-Mobility.pdf

[551] IEA. Well to wheels greenhouse gas emissions for cars by power trains. (2021). https://www.iea.org/data-and-statistics/charts/well-

to-wheels-greenhouse-gas-emissions-for-cars-by-powertrains
Note: Life cycle GHG emissions were estimated by adding vehicle manufacturing emissions to well-to-wheels emissions. This information was available from another IEA report: Comparative life cycle GHG emissions from a midsize BEV and ICE vehicle (2021). https://www.iea.org/data-and-statistics/charts/comparative-life-cycle-greenhouse-gas-emissions-of-a-mid-size-bev-and-ice-vehicle

[552] T. V. Choudhary. Critical Comparison of Low Carbon Technologies (October 2020). https://www.amazon.com/dp/B08LP8TRLP

[553] MIT Energy Initiative, Insights into Future Mobility (November 2019): http://energy.mit.edu/wp-content/uploads/2019/11/Insights-into-Future-Mobility.pdf

[554] Argonne National Laboratory (April 2021): Comprehensive total cost of ownership calculation for vehicles with different size classes and power trains. https://publications.anl.gov/anlpubs/2021/05/167399.pdf

[555] T. V. Choudhary. Critical Comparison of Low Carbon Technologies (October 2020). This reference book provides an estimate for average prices based on representative vehicles. https://www.amazon.com/dp/B08LP8TRLP

[556] U.S. DOE, Office of Energy Efficiency and Renewable Energy. Developing infrastructure to charge plug -in electric vehicles. https://afdc.energy.gov/fuels/electricity_infrastructure.html#dc

[557] Note: Over 80% of charging stations in the United States are Level 2 (25 to 70 miles of charging per hour). https://afdc.energy.gov/fuels/electricity_locations.html#/find/nearest?fuel=ELEC

[558] Note: Results from the U.S. Department of Transportation study provide quality mass transit data. The study showed that for bus transit, the CO_2 emissions were 82 gCO_2 per passenger mile for 100% (maximum) occupancy of the bus. For commuter train, the CO_2 emissions were 45 gCO_2 per passenger mile for 100% (maximum) occupancy. The study also showed an average CO_2 emission of 327 gCO_2/per passenger mile for private auto trips based on the observed average occupancy of 1.4 passengers per private auto trip for work and other general reasons. The 60+% reduction number for mass transit corresponds to 70% vehicle occupancy.

[559] T. V. Choudhary. Critical Comparison of Low Carbon Technologies (October 2020). This reference book provides an estimate of average costs for different vehicle technologies. https://www.amazon.com/dp/B08LP8TRLP

[560] Note: Apart from using subsidies, examples of other incorrect approaches include comparing vehicles with very dissimilar travel ranges and using narrow time periods for fuel prices. The minimum travel range should be 250 miles because conventional vehicles have a travel range of 400 miles. Vehicles with the same trim should be considered to minimize impact from external factors. Also, certain studies use fuel prices from a narrow time range. Considering that fuel prices fluctuate substantially, the most reasonable approach is to use average fuel price from the last ten to fifteen years.

[561] MIT Energy Initiative, Insights into Future Mobility (November 2019): http://energy.mit.edu/wp-content/uploads/2019/11/Insights-into-Future-Mobility.pdf

[562] T. V. Choudhary. Critical Comparison of Low Carbon Technologies (October 2020). This reference book provides an estimate of average costs for different vehicle technologies. https://www.amazon.com/dp/B08LP8TRLP

[563] Argonne National Laboratory (April 2021): Comprehensive total cost of ownership calculation for vehicles with different size classes and power trains. https://publications.anl.gov/anlpubs/2021/05/167399.pdf

[564] McKinsey & Company Special Report (January 2022): The net zero transition. What it would cost? What it could bring? https://www.mckinsey.com/business-functions/sustainability/our-insights/the-economic-transformation-what-would-change-in-the-net-zero-transition

[565] U.S. EPA: https://www.epa.gov/sites/default/files/2016-11/documents/vw-faqs-app-c-final-11-18-16.pdf https://www.epa.gov/newsreleases/epa-finalizes-greenhouse-gas-standards-passenger-vehicles-paving-way-zero-emissions

[566] Final report for BEUC (The European Consumer Organization)- April 2021: Electric cars: Calculating the total cost of ownership for consumers. https://www.beuc.eu/publications/beuc-x-2021-039_electric_cars_calculating_the_total_cost_of_ownership_for_consumers.pdf

[567] Federal planning Bureau (December 2019): Total cost of ownership of electric cars compared to diesel and gasoline cars in Belgium. https://www.plan.be/uploaded/documents/202001131009220.REP_VHSTCOBEV_12036.pdf

[568] Hsien and Green (MIT): Transition to electric vehicles in China: Implications for total cost of ownership and societal costs. SAE, J. STEEP, 1, 87 (2020). https://doi.org/10.4271/13-01-02-0005

[569] BP Statistical review of World energy 2021. https://www.bp.com/content/dam/bp/business-sites/en/global/corporate/pdfs/energy-economics/statistical-review/bp-stats-review-2021-full-report.pdf

[570] The International council on clean transportation (ICCT). A global comparison of the lifecycle gas emissions of combustion engine and electric passenger cars. https://theicct.org/sites/default/files/publications/Global-LCA-passenger-cars-jul2021_0.pdf

[571] Note the ICCT reference appears to be biased towards BEVs. It suggests that the current reductions in the U.S. are about 62%. As discussed in the text, an apples-to-apples comparison (i.e., for same model and trims based on official U.S. Economy.gov data), shows only about a 50% reduction for small-to-mid-sized vehicles. Also, according to the ICCT report, hybrid vehicles reduce GHG emissions by 20%. Examples in the text show a 25% to 30% reduction when vehicles are compared for the same model and trim. The ICCT report assumes an average of vehicles sold in the U.S., which skews towards large vehicles. This is incorrect from a realistic comparison viewpoint, considering that BEVs today are mostly concentrated in the low to mid-sized vehicles. Correspondingly, the ICCT report uses a higher value for GHG emissions from ICE cars. A realistic comparison should only consider the small-to-midrange vehicles, because most sales of BEVs belong to that category. This is not expected to change drastically at-least for the next several years.

[572] National Renewable Energy Laboratory (October 2021) Report. There is no place like home: residential parking, electrical access and implications for the future of electric vehicle charging infrastructure. https://www.nrel.gov/docs/fy22osti/81065.pdf

[573] The International Council on Clean Transportation White Paper (July 2021): Charging up America: Assessing the growing needs of U.S. infrastructure through 2030.

https://theicct.org/publication/charging-up-america-assessing-the-growing-need-for-u-s-charging-infrastructure-through-2030/

[574] U.S. Alternative fuels data center. Alternative fueling station locator https://afdc.energy.gov/stations/#/analyze

[575] The International Council on Clean Transportation White Paper (July 2021): Charging up America: Assessing the growing needs of U.S. infrastructure through 2030. https://theicct.org/publication/charging-up-america-assessing-the-growing-need-for-u-s-charging-infrastructure-through-2030/

[576] U.S. Bureau of Transportation statistics: https://www.bts.gov/content/number-us-aircraft-vehicles-vessels-and-other-conveyances

[577] U.S. Alternative fuels data center. Developing infrastructure to charge plugin electric vehicles. https://afdc.energy.gov/fuels/electricity_infrastructure.html#dc

[578] U.S. Alternative fuels data center. Alternative fueling station locator https://afdc.energy.gov/stations/#/analyze?country=US&fuel=ELEC

[579] ACEA. E-mobility: Only 1 in 9 charging points in EU is fast. https://www.acea.auto/press-release/e-mobility-only-1-in-9-charging-points-in-eu-is-fast/

[580] Rocky Mountain Institute: Reducing EV charging infrastructure costs. https://rmi.org/wp-content/uploads/2020/01/RMI-EV-Charging-Infrastructure-Costs.pdf

[581] Mach1 Road assistance services. Average costs of using car charging stations. https://www.mach1services.com/costs-of-using-car-charging-stations/

[582] Electrify America: Pricing and plans for EV charging. https://www.electrifyamerica.com/pricing/

[583] EVgo fast charging. EVgo fast charging pricing. https://www.evgo.com/pricing/

[584] Shell: How much does Shell recharge cost? https://support.shell.com/hc/en-gb/articles/115002988472-How-much-does-Shell-Recharge-cost-

[585] EnelX: The ultimate guide to electric vehicle public charging. https://evcharging.enelx.com/resources/blog/579-the-ultimate-guide-to-electric-vehicle-public-charging-pricing

[586] MIT Energy Initiative, Insights into Future Mobility (November 2019): http://energy.mit.edu/wp-content/uploads/2019/11/Insights-into-Future-Mobility.pdf

[587] T. V. Choudhary. Critical Comparison of Low Carbon Technologies (October 2020). This reference book provides an estimate of average costs for different vehicle technologies. https://www.amazon.com/dp/B08LP8TRLP

[588] Argonne National Laboratory (April 2021): Comprehensive total cost of ownership calculation for vehicles with different size classes and power trains. https://publications.anl.gov/anlpubs/2021/05/167399.pdf

[589] Anderson Economic Group (October 2021 Report): Real world electric fueling costs might surprise new drivers. https://www.andersoneconomicgroup.com/real-world-electric-vehicle-fueling-costs-may-surprise-new-ev-drivers/

[590] Fueleconomy.gov. https://www.fueleconomy.gov/feg/Find.do?action=sbsSelect

[591] Note: Nissan Leaf S Plus (230 miles travel range) is substantially more expensive than Nissan Leaf S (150 miles travel range). Nissan Website. 2021 Nissan LEAF https://www.nissanusa.com/vehicles/electric-cars/leaf.html

[592] Tesla website. Tesla Model 3. https://www.tesla.com/model3/design#overview

[593] IPCC Report: Carbon dioxide capture and storage (2005). https://www.ipcc.ch/site/assets/uploads/2018/03/srccs_wholereport-1.pdf

[594] U.S. National oceanic and atmospheric administration, Earth System Research Laboratories: Trends in atmospheric CO_2. https://www.esrl.noaa.gov/gmd/ccgg/trends/

[595] IPCC Special Report: Global warming of 1.5^{O} C. https://www.ipcc.ch/sr15/chapter/chapter-4/

[596] U.S. Department of Energy. Carbon capture opportunities for natural gas power systems. https://www.energy.gov/sites/prod/files/2017/01/f34/Carbon%20Capture%20Opportunities%20for%20Natural%20Gas%20Fired%20Power%20Systems_0.pdf

[597] IPCC Report: Carbon dioxide capture and storage (2005). https://www.ipcc.ch/site/assets/uploads/2018/03/srccs_wholereport-1.pdf

[598] Biodiversity. Chapter 2 by Paul Ehrlich. The cause of diversity losses and consequences. https://www.ncbi.nlm.nih.gov/books/NBK219310/

[599] Extensive research involves robust scientific principles and large amount of quality data.

[600] IPCC Report: Climate Change 2021, Physical Basis. Contribution of working group I to the sixth assessment report. https://www.ipcc.ch/report/ar6/wg1/#FullReport

[601] Note: Urgent attention means initiating actions as soon as possible, which are driven by science and practical considerations.

[602] Center of Disease Control and Prevention. Smoking and Tobacco use. Health effects. https://www.cdc.gov/tobacco/basic_information/health_effects/index.htm

[603] World Health organization factsheets: Tobacco. https://www.who.int/news-room/fact-sheets/detail/tobacco

[604] World Meteorological Organization Report # 1267 (2021): WMO Atlas of mortality and economic losses from weather, water and climate extremes, 1970-2019. https://library.wmo.int/index.php?lvl=notice_display&id=21930#.YvFKgi-B3b0

[605] Netherlands Environmental Assessment Energy, December 2020 Report "Trends in global CO_2 and total greenhouse gas emissions; 2020 report". https://www.pbl.nl/en/publications/trends-in-global-co2-and-total-greenhouse-gas-emissions-2020-report

[606] World Meteorological Organization Report # 1267 (2021): WMO Atlas of mortality and economic losses from weather, water and climate extremes, 1970-2019. https://library.wmo.int/index.php?lvl=notice_display&id=21930#.YvFKgi-B3b0

[607] Note: There is a general trend of increasing disasters. But there are exceptions. For example, the 2000s had more disasters than the 2010s.

[608] United Nations. Climate action. Disasters, resilience and land management. https://www.un.org/en/climatechange/climate-solutions/disasters-resilience-land-management

[609] World Meteorological Organization Report # 1267 (2021): WMO Atlas of mortality and economic losses from weather, water and climate extremes, 1970-2019.

https://library.wmo.int/index.php?lvl=notice_display&id=21930#.YvFKgi-B3b0

[610] IPCC Report: Climate Change 2021, Physical Basis. Contribution of working group I to the sixth assessment report. https://www.ipcc.ch/report/ar6/wg1/#FullReport

[611] World Resource Institute. The New Climate Economy Report. https://newclimateeconomy.report/2018/wp-content/uploads/sites/6/2018/09/NCE_2018_FULL-REPORT.pdf

[612] Steve Koonin. Unsettled: What climate science tells us, What it doesn't, Why it matters. April (2021).

[613] IPCC reports and the tens of thousands of references therein. https://www.ipcc.ch/reports/

[614] IPCC Sixth Assessment Report (AR6). The Physical Science Basis. https://www.ipcc.ch/assessment-report/ar6/

[615] IPCC Sixth Assessment Report (AR6). The Physical Science Basis (2021). And references therein. https://www.ipcc.ch/assessment-report/ar6/

[616] IPCC Sixth Assessment Report (AR6). Impacts, Adaptation and Vulnerability (2022). And references therein. https://www.ipcc.ch/assessment-report/ar6/

[617] World Bank data. GDP growth (annual %) World. https://data.worldbank.org/indicator/NY.GDP.MKTP.KD.ZG?locations=1W

[618] International Monetary Fund. World Economic Outlook, October 2019. https://www.imf.org/en/Publications/WEO/Issues/2019/10/01/world-economic-outlook-october-2019

[619] Note: Percentage point is a numerical difference between two percentages. Percent (%) is the ratio between two numbers.

[620] Congressional Budget Office report, September 2020. CBO's projection of the effect of climate change on U.S. Economic Output. https://www.cbo.gov/publication/56505

[621] World Bank data. GDP growth (annual %) United States. https://data.worldbank.org/indicator/NY.GDP.MKTP.KD.ZG?locations=US

[622] Note: Percentage point is a numerical difference between two percentages. Percent (%) is the ratio between two numbers.

[623] Swiss Re Institute report. April 2021. The economics of climate change. No action not an option. https://www.swissre.com/institute/research/topics-and-risk-

dialogues/climate-and-natural-catastrophe-risk/expertise-publication-economics-of-climate-change.html

[624] Note: Swiss Re report defines the unknown climate change issues as the unknown unknowns.

[625] Swiss Re Institute report. April 2021. The economics of climate change. No action not an option. https://www.swissre.com/institute/research/topics-and-risk-dialogues/climate-and-natural-catastrophe-risk/expertise-publication-economics-of-climate-change.html

[626] NGFS climate scenarios for central banks and supervisors. https://www.ngfs.net/sites/default/files/media/2021/08/27/ngfs_climate_scenarios_phase2_june2021.pdf

[627] Note: The economic impacts were only modelled out to 2050. For the 2100 year estimate the study assumed only physical risk impacts. They estimated the global GDP to be lower by 11% in 2100 for the current policies compared to net zero 2050 policies.

[628] Note: It is perhaps alright to discuss the worst possible impact. But it is critical to mention the probability for such outcomes. For example, a global nuclear war or pandemic is far more likely. It is misleading to discuss the massive uncertainty of potential unknowns with such certainty.

[629] Note: the dollar values are comparable across countries because they are based on purchasing power parity (PPP).

[630] UN Sustainable Development Goals. https://www.un.org/sustainabledevelopment/sustainable-development-goals/

[631] Note: According to MIT living wage calculator, it takes over 50$/day to get by in the United States. https://livingwage.mit.edu

[632] Note: Recall, for the purpose of this discussion, the economically distressed category is defined as the fraction of the population that lives under 10$ (2011 PPP)/day.

[633] The World Bank. PovcalNet. http://iresearch.worldbank.org/PovcalNet/povDuplicateWB.aspx

[634] United Nations Development Programme Report. People's climate vote: Results. https://www.undp.org/publications/peoples-climate-vote

[635] Note: Over one million people from fifty countries participated in the survey. The survey was undertaken via adverts in popular mobile gaming apps. Considering that the population in the economically distressed category is expected to have limited access to mobile

gaming apps, the survey very likely undercounted responses from this category.

[636] Note: According to the report, average of 64% of the people surveyed believed that climate change was an emergency. Of this subset, 59% agreed that everything necessary must be done urgently. Thus, percent of total people that agreed that everything necessary must be done urgently = 64*59% = 38%

[637] Note: According to a 2021 U.S. based Pew Research Center survey, climate change was found to be a top personal concern for only 31% of the adults. Reference: https://www.pewresearch.org/fact-tank/2021/05/26/key-findings-how-americans-attitudes-about-climate-change-differ-by-generation-party-and-other-factors/ & the United Nations Development Programme Report discussed earlier.

[638] United Nations Development Programme Report. People's climate vote: Results. https://www.undp.org/publications/peoples-climate-vote

[639] United Nations Development Programme Report. People's climate vote: Results. https://www.undp.org/publications/peoples-climate-vote

[640] Pew Research Center Report (September 2021) focused on advanced economies. https://www.pewresearch.org/global/2021/09/14/in-response-to-climate-change-citizens-in-advanced-economies-are-willing-to-alter-how-they-live-and-work/

[641] Note: Climate change is the driver for the global push towards a low-carbon transition. If CO_2 impact on climate change had not been recognized, there would be no emphasis on a low-carbon transition.

[642] American Institute of Physics. https://history.aip.org/climate/summary.htm

[643] Data from "Carbon Dioxide Information Analysis Center". T. Boden, D. Andres, Oakridge National Laboratory. https://cdiac.ess-dive.lbl.gov/ftp/ndp030/global.1751_2014.ems

[644] Note: For a size context, a total of 10 billion tons of plastics have been produced to-date. UN Environment. Global Chemicals Outlook II, From Legacies to Innovative solutions (2019). https://wedocs.unep.org/bitstream/handle/20.500.11822/27651/GCOII_synth.pdf?sequence=1&isAllowed=y

[645] G.S. Calendar, Quarterly Journal of the Royal Meteorological Society. April 1938. The artificial production of CO_2 and its

influence on temperature. https://rmets.onlinelibrary.wiley.com/doi/abs/10.1002/qj.49706427503

[646] J.R. Fleming, Eos, Transactions American Geophysical Union 79, 405 (1998).

[647] United States President's Science Advisory Committee (1965), Restoring the quality of our environment. https://www.worldcat.org/title/restoring-the-quality-of-our-environment-report-of-the-environmental-pollution-panel-of-the-presidents-science-advisory-committee/oclc/562799 **Note:** The only recommendation in this report by the esteemed panel of scientists related to this topic was to continue the precise measurements of CO_2 and temperature at different heights in the stratosphere (Page 26). The level of understanding of severity can be evaluated by considering the set of recommendations in this study. Lack of any recommendation to reduce CO_2 is the best evidence for an inadequate understanding about the severity of the effect from CO_2.

[648] Note: Although a warning about significant potential climate impacts was issued, the lack of adequate understanding was also acknowledged in the 1965 report. From page 114: "Even today, we cannot make a useful prediction concerning the nature or the magnitude of the possible climatic effects"

[649] Data from "Carbon Dioxide Information Analysis Center". T. Boden, D. Andres, Oakridge National Laboratory. https://cdiac.ess-dive.lbl.gov/ftp/ndp030/global.1751_2014.ems

[650] U.S. National Academy of Science Report: Understanding Climate Change (1975). https://archive.org/details/understandingcli00unit/mode/2up

[651] Data from "Carbon Dioxide Information Analysis Center". T. Boden, D. Andres, Oakridge National Laboratory. https://cdiac.ess-dive.lbl.gov/ftp/ndp030/global.1751_2014.ems

[652] IPCC history. https://www.ipcc.ch/about/history/

[653] Climate Change: The 1990 and 1992 IPCC Assessments. https://www.ipcc.ch/site/assets/uploads/2018/05/ipcc_90_92_assessments_far_overview.pdf

[654] IPCC reports. https://www.ipcc.ch/reports/

[655] Note: The temperature rise or warming was so small during the first few decades of the twentieth century, that it would have been

impossible to reasonably accept the theory that there was an impact from fossil fuel technologies.

[656] Note: This is being generous. Trends in temperature were only obvious by the year 1980. https://climate.nasa.gov/vital-signs/global-temperature/ For example, there was no discernible rise in global temperature from 1940 to 1975. So, 1980 could also be considered as the year that fossil fuel technologies reach the first minimum level of implementation. Recall, scientific basis requires reasonable evidence.

[657] Climate Change: The 1990 and 1992 IPCC Assessments. https://www.ipcc.ch/site/assets/uploads/2018/05/ipcc_90_92_assessments_far_overview.pdf

[658] Data from "Carbon Dioxide Information Analysis Center". T. Boden, D. Andres, Oakridge National Laboratory. https://cdiac.ess-dive.lbl.gov/ftp/ndp030/global.1751_2014.ems

[659] IPCC reports. https://www.ipcc.ch/reports/

[660] American Institute of Physics website. https://history.aip.org/climate/co2.htm

[661] G.S. Calendar, Quarterly Journal of the Royal Meteorological Society. April 1938. The artificial production of CO_2 and its influence on temperature. https://rmets.onlinelibrary.wiley.com/doi/abs/10.1002/qj.49706427503

[662] BBC News online. https://www.bbc.com/news/uk-england-norfolk-22283372

[663] United States President's Science Advisory Committee (1965), Restoring the quality of our environment. https://www.worldcat.org/title/restoring-the-quality-of-our-environment-report-of-the-environmental-pollution-panel-of-the-presidents-science-advisory-committee/oclc/562799

[664] Our World in Data. Global direct energy consumption. https://ourworldindata.org/grapher/global-primary-energy?country=~OWID_WRL

[665] Note: The data can be obtained from following two references. "Carbon Dioxide Information Analysis Center". T. Boden, D. Andres, Oakridge National Laboratory. https://cdiac.ess-dive.lbl.gov/ftp/ndp030/global.1751_2014.ems and BP Statistical review of World energy 2021. https://www.bp.com/content/dam/bp/business-

sites/en/global/corporate/pdfs/energy-economics/statistical-review/bp-stats-review-2021-full-report.pdf

[666] Our World in Data. Global direct energy consumption. https://ourworldindata.org/grapher/global-primary-energy?country=~OWID_WRL

[667] U.S. EIA: Natural gas explained. https://www.eia.gov/energyexplained/natural-gas/

[668] Note: Solar and wind power will require supporting technologies such as energy storage to produce on-demand electricity. The impact from any such supporting technology is included in the discussion.

[669] European Environment agency- Human activities https://www.eea.europa.eu/publications/92-827-5122-8/page011.html

[670] Wikipedia: Human impact on the environment. https://en.wikipedia.org/wiki/Human_impact_on_the_environment

[671] Note: Solar and wind power require supporting technologies such as energy storage to produce on-demand electricity. The impact from any such supporting technology is included in the discussion.

[672] Adani Renewables. 648 MW, Kamuthi Tamil Nadu. https://www.adanigreenenergy.com/solar-power/Kamuthi

[673] Environmental Justice Atlas. Kamuthi Solar project. https://ejatlas.org/conflict/kamuthi-solar-power-project-648-mw-tamil-nadu-india

[674] Oakridge National Laboratory (2017): Environmental quality and U.S. power sector- air quality, water quality, land use and environmental justice. https://www.energy.gov/sites/prod/files/2017/01/f34/Environment%20Baseline%20Vol.%202--Environmental%20Quality%20and%20the%20U.S.%20Power%20Sector--Air%20Quality%2C%20Water%20Quality%2C%20Land%20Use%2C%20and%20Environmental%20Justice.pdf

[675] The new Indian Express. Adani solar plant guzzles illegal water from drought hit Tamil Nadu. https://www.newindianexpress.com/states/tamil-nadu/2017/jun/06/adani-solar-plant-guzzles-illegal-fresh-water-in-drought-hit-tamil-nadu-1613326.html

[676] Note: At a wide-scale implementation, substantial amount of energy storage will be required. While wind power will help in

decreasing the required energy storage, substantial amount of energy storage will still be required because of daily and seasonal variation of wind and solar.

[677] World Bank group: Minerals for climate action. The minerals intensity of clean energy transition. https://pubdocs.worldbank.org/en/961711588875536384/Minerals-for-Climate-Action-The-Mineral-Intensity-of-the-Clean-Energy-Transition.pdf

[678] The STM report, 2015. An overview of scientific and scholarly journal publishing. https://www.stm-assoc.org/2015_02_20_STM_Report_2015.pdf

[679] Note: The only exception would be if the conclusions in the isolated research article were based on basic scientific principles that cannot be refuted.

[680] Nature. Climate simulations recognize the hot model problem. https://www.nature.com/articles/d41586-022-01192-2

[681] Science. Use of too hot climate models exaggerates impact of global warming. https://www.science.org/content/article/use-too-hot-climate-models-exaggerates-impacts-global-warming

[682] IPCC Report: Climate Change 2021, Physical Basis. Contribution of working group I to the sixth assessment report. https://www.ipcc.ch/report/ar6/wg1/#FullReport

[683] Note: The only exception is for research conclusions that are exclusively based on basic scientific principles.

[684] PBL Netherlands Environmental Assessment Agency 2020 Report. Trends in global CO_2 and GHG emissions. https://www.pbl.nl/en/publications/trends-in-global-co2-and-total-greenhouse-gas-emissions-2020-report

[685] IEA Technology Report. Global energy review: CO_2 emissions in 2021. https://www.iea.org/reports/global-energy-review-co2-emissions-in-2021-2

[686] Climate Watch: Historical GHG emissions. https://www.climatewatchdata.org/ghg-emissions?breakBy=regions&end_year=2018&gases=ch4®ions=WORLD%2CWORLD§ors=electricity-heat&start_year=1990

[687] World Resources Institute: 4 charts explain greenhouse gas emissions by countries and sectors? https://www.wri.org/insights/4-charts-explain-greenhouse-gas-emissions-countries-and-sectors

[688] Note: Examples of agriculture include crop cultivation and livestock. Examples of direct industrial processes include production of cement and chemicals. Examples of waste include landfills and wastewater.

[689] Our World in Data. CO_2 emissions. https://ourworldindata.org/co2-emissions

[690] United Nations. Climate action. https://www.un.org/en/climatechange/net-zero-coalition

[691] PBL Netherlands Environmental Assessment Agency 2020 Report. Trends in global CO_2 and GHG emissions. https://www.pbl.nl/en/publications/trends-in-global-co2-and-total-greenhouse-gas-emissions-2020-report

[692] Note: For the purpose of this discussion, an established electrification technology is one that has been substantially deployed. Substantial is in terms of absolute deployment and not relative to conventional fossil fuel technologies.

[693] IEA: Transport. Improving the sustainability of passengers and freight transport. https://www.iea.org/topics/transport

[694] McKinsey & Company, May 2020. Plugging in: What electrification can do for the industry. https://www.mckinsey.com/industries/electric-power-and-natural-gas/our-insights/plugging-in-what-electrification-can-do-for-industry

[695] IEA, September 2020. The challenges of reaching zero emissions in heavy industry. https://www.iea.org/articles/the-challenge-of-reaching-zero-emissions-in-heavy-industry

[696] McKinsey & Company Special Report (January 2022): The net zero transition. What it would cost? What it could bring? https://www.mckinsey.com/business-functions/sustainability/our-insights/the-economic-transformation-what-would-change-in-the-net-zero-transition

[697] Note: Estimating the cost for a global transition is extremely challenging because of its vast scope. Therefore, tremendous expertise is required for a reasonable estimate. Such expertise resides in very few organizations, for example in global consulting firms such as McKinsey & Company. However, even estimations from experts are expected to have significant uncertainty. It is wise to ignore estimations from groups that do not have real-world expertise with very large scope projects, for example academia or research organizations.

[698] McKinsey & Company Special Report (January 2022): The net zero transition. What it would cost? What it could bring? https://www.mckinsey.com/business-functions/sustainability/our-insights/the-economic-transformation-what-would-change-in-the-net-zero-transition

[699] Note: For example, the study estimates that about 2 trillion dollars of assets will be stranded in the power sector alone. Early retirements of a massive number of productive assets will lead to bankruptcies, credit defaults and has the potential to rock the global financial system. Also, without redundancy in energy supply systems, the occasional energy supply disruptions caused by the energy transition will cause chaos. Unfortunately, including adequate redundancy in energy systems will result in enormous additional costs.

[700] U.S. Geological Survey: Critical Minerals. https://www.usgs.gov/science/critical-minerals

[701] The World Bank Group (2020 report). Minerals for climate action: the minerals intensity of clean energy transition. https://pubdocs.worldbank.org/en/961711588875536384/Minerals-for-Climate-Action-The-Mineral-Intensity-of-the-Clean-Energy-Transition.pdf

[702] IEA Report, revised March 2022. The role of critical minerals on clean energy transitions. https://www.iea.org/reports/the-role-of-critical-minerals-in-clean-energy-transitions

[703] BP Statistical review of World energy 2021. https://www.bp.com/content/dam/bp/business-sites/en/global/corporate/pdfs/energy-economics/statistical-review/bp-stats-review-2021-full-report.pdf

[704] International Energy Agency: Tracking Transport, Electric Vehicles, Tracking Report November 2021. https://www.iea.org/reports/electric-vehicles

[705] IEA Report, revised March 2022. The role of critical minerals on clean energy transitions. https://www.iea.org/reports/the-role-of-critical-minerals-in-clean-energy-transitions

[706] A Report by the White House. Building resilient supply chains, revitalizing American manufacturing, and fostering broad based growth. https://www.whitehouse.gov/wp-content/uploads/2021/06/100-day-supply-chain-review-report.pdf

[707] A Report by the White House. Building resilient supply chains, revitalizing American manufacturing, and fostering broad based

growth. https://www.whitehouse.gov/wp-content/uploads/2021/06/100-day-supply-chain-review-report.pdf

[708] IEA Report, revised March 2022. The role of critical minerals on clean energy transitions. https://www.iea.org/reports/the-role-of-critical-minerals-in-clean-energy-transitions

[709] The World Bank Group (2020 report). Minerals for climate action: the minerals intensity of clean energy transition. https://pubdocs.worldbank.org/en/961711588875536384/Minerals-for-Climate-Action-The-Mineral-Intensity-of-the-Clean-Energy-Transition.pdf

[710] U.S. Geological Survey, 2021. Minerals Commodity Summaries 2021. https://pubs.usgs.gov/periodicals/mcs2021/mcs2021.pdf

[711] OECD Environment Policy paper No. 12 (Sept. 2018). Improving plastics management: trends, policy, and the role of international cooperation and trade. https://www.oecd.org/environment/waste/policy-highlights-improving-plastics-management.pdf

[712] OECD.org Plastic pollution is growing relentlessly as waste management and recycling falls short, says OECD. https://www.oecd.org/environment/plastic-pollution-is-growing-relentlessly-as-waste-management-and-recycling-fall-short.htm

[713] U.S. EPA. Facts and figures about materials, waste and recycling. https://www.epa.gov/facts-and-figures-about-materials-waste-and-recycling/plastics-material-specific-data

[714] United National Environment Programme, 2019 press release. Time to seize opportunity, tackle challenge of e-waste. https://www.unep.org/news-and-stories/press-release/un-report-time-seize-opportunity-tackle-challenge-e-waste

[715] MIT Technology Review, Aug. 2021. Solar panels are a pain to recycle. These companies are trying to fix that. https://www.technologyreview.com/2021/08/19/1032215/solar-panels-recycling/

[716] Note: Because of the large size, the wind turbine blades require large amounts of raw materials. For reference, a wind turbine blade can be longer than a Boeing 777 wing.

[717] IEA Wind Task 45. Enabling the recycling of wind turbine blades. https://iea-wind.org/task45/

[718] U.S. EIA, Wind Energy and the Environment. https://www.eia.gov/energyexplained/wind/wind-energy-and-the-environment.php

[719] The World Bank Group (2020 report). Minerals for climate action: the minerals intensity of clean energy transition. https://pubdocs.worldbank.org/en/961711588875536384/Minerals-for-Climate-Action-The-Mineral-Intensity-of-the-Clean-Energy-Transition.pdf

[720] Note: GDP is the monetary value of all goods and services made within a country during a specific period.

[721] World Bank Data: GDP per capita PPP (constant 2017 International dollar). PPP = purchasing power parity. https://data.worldbank.org/indicator/NY.GDP.PCAP.PP.KD Note: Such data allows more accurate comparison of countries by accounting for their different buying powers. For example, a dollar can buy more food in India than in the United States.

[722] Energy data was estimated from U.S. EIA: https://www.eia.gov/international/data/world/total-energy/total-energy-consumption

[723] Population data from World Bank Data. https://data.worldbank.org/indicator/SP.POP.TOTL

[724] Food Administration and organization of the United Nations. The energy and agricultural nexus. Environmental and natural resources working paper No. 4 (2000).

[725] United Nations. Climate Action. https://www.un.org/en/climatechange/net-zero-coalition

[726] Our World in Data. Global direct primary consumption. https://ourworldindata.org/grapher/global-primary-energy?country=~OWID_WRL

[727] International Energy Agency Report (2021). Net zero by 2050. A roadmap for the energy sector. https://www.iea.org/reports/net-zero-by-2050 Note: low carbon technologies include fossil fuel technologies with CCS.

[728] Note: Average annual energy consumption during previous energy transition is average of the annual energy consumed from 1860 to 1960. The data is obtained from Our World in Data.

[729] Our World in Data. Global direct primary consumption. https://ourworldindata.org/grapher/global-primary-energy?country=~OWID_WRL

[730] U.S. Geological Survey, 2021. Minerals Commodity Summaries 2021. https://pubs.usgs.gov/periodicals/mcs2021/mcs2021.pdf

[731] IEA Report, revised March 2022. The role of critical minerals on clean energy transitions. https://www.iea.org/reports/the-role-of-critical-minerals-in-clean-energy-transitions

[732] Note: The large swings or cycles in energy prices over the decades are well documented. These cycles or swings have been caused by disruptions in energy supply because of various reasons such as regional conflicts, disasters, and over or under capital allocation for energy production and processing.

[733] Note: Currently, low-carbon power technologies are substantially more costly relative to either coal power or natural gas power when compared on an apples-to-apples basis, i.e., for 24X7 on-demand electricity production. In other words, solar and wind power costs cannot be compared without including associated costs of energy storage and/or overbuilding and/or extensive transmission grids. However, costs can be directly compared for low-carbon technologies such as nuclear power because it can provide 24X7 on-demand electricity.

[734] U.S. Energy Information Administration: Levelized cost of new generation resources in the annual energy outlook 2021. https://www.eia.gov/outlooks/aeo/pdf/electricity_generation.pdf

[735] OECD NEA report (2019): System costs with high share of nuclear energy and renewables. https://www.oecd-nea.org/jcms/pl_15000/the-costs-of-decarbonisation-system-costs-with-high-shares-of-nuclear-and-renewables?details=true

[736] International Energy Agency (2021): World Energy model. https://www.iea.org/reports/world-energy-model/techno-economic-inputs

[737] Note: higher human activity and resources for the same end-product (24X7 electricity) equals higher cost.

[738] Note: Publications include articles and books.

[739] Note: Most climate activists discuss fossil fuels only to highlight their detrimental impacts. They believe that fossil fuels should never have been used and therefore cannot bring themselves to consider using fossil fuels as a reference point. In other words, their strong negative emotions are responsible for their inability to use the robust reference point approach to understand the severity of the challenges associated with the low-carbon energy transition.

[740] World Bank. Fossil fuel energy consumption, percent of total. https://data.worldbank.org/indicator/EG.USE.COMM.FO.ZS

[741] IPCC reports author list and publications cited in the reports. https://apps.ipcc.ch/report/authors/ Note: Academia includes personnel from University and National Laboratories.

[742] Note: Academicians are best equipped to discuss climate change impacts because the discussions are based on climate change science.

[743] Note: Practical considerations and knowledge about the energy industry are crucial for developing a robust path forward for the low-carbon transition. Most academicians do not have this required expertise. This has led to unrealistic optimism discussed earlier.

[744] Note: Being a mature industry, relatively few academic researchers are involved with fossil fuel research. Also, the focus on academic researchers in fundamental research and less on practical considerations.

[745] Note: The energy industry has important commonalities irrespective of the energy source such as massive infrastructures or physical assets, massive resource (personnel and materials) requirements, and maintaining the energy supply to robustly meet the enormous energy demand. Therefore, energy experience in the oil and gas industry is generically transferrable to low-carbon energy.

[746] Note: This should be true for most of the energy experts.

[747] Note: While this is a valid concern, there are ways to get around this. For example, retired personnel from the industry who do not have the same financial interests/concerns (e.g., job security concerns, etc.) as current employees can be sought out.

[748] Note: Efficiently addressing climate change means addressing climate change both swiftly and sustainably.

[749] Note: For details, see chapters 1-3 & 9.

[750] Note: For details see chapter 3.

[751] Note: For details see chapter 5.

[752] Note: Most people, even in advanced economies, are not willing to significantly increase their energy bills in support of climate change mitigation. For example, a recent survey showed that 52% of Americans supported an additional payment of 1$/month for energy costs, while 28% percent opposed it. About 31% supported an additional payment of 100$/month, while 52% opposed it. https://epic.uchicago.edu/wp-content/uploads/2021/10/8784-EPIC-Final-Topline.Final_.pdf

[753] Note: In countries with low-prices natural gas, this is a very cost-effective option. Natural gas power plants can reduce greenhouse gas emissions from coal power plants by 50%, whereas most low-carbon power technologies can achieve greater than 90% reduction. But potential exists for further greenhouse gas reductions from natural gas power plants by including a capture and storage component. Thus, it is an excellent intermediate-term solution.

[754] Note: Otherwise, the developing countries are likely to deploy coal power plants.

[755] Note: Policies should focus on using shuttle vans for transporting passengers. This will ensure excellent area coverage and low wait-times.

[756] Note: This means that climate enthusiasts should purchase BEVs and bear the additional cost. Tax money must not be used for this purpose. This should not be a problem for the climate enthusiasts because they are extremely passionate about mitigating climate change. Moreover, most climate enthusiasts are affluent and should be easily able to afford the extra cost.

[757] Note: An electrical grid with a low carbon intensity means having less than 150 gCO_2 produced per kWh of electricity. For reference, the value for Canada is less than 120 gCO_2/kWh. https://www.cer-rec.gc.ca/en/data-analysis/canada-energy-future/2021/towards-net-zero.html United States has a carbon intensity of electricity generation of about 380 gCO_2 per kWh, while Australia and India have a carbon intensity that is above 650 gCO_2 per kWh. The carbon intensity of electricity generation in China is above 500 gCO_2 per kWh.

[758] Note: McKinsey and company estimated that 200 million new jobs will be created because of the low-carbon transition. The United States consumes 17% of the global energy.

[759] Note: Climate adaptation is crucial for avoiding substantial life and property and livelihood loss **today**. It cannot be delayed for addressing a future crisis.

[760] Note: This policy ties well into the last point discussed in the previous section, i.e., discussions related to Requirement 2.

[761] Note: Travel in private light duty vehicles requires substantially more vehicle miles and larger number of vehicles compared to mass transit. Thus, policy support for personal light duty vehicles

is equivalent to encouraging higher energy consumption and larger exploitation of resources.

[762] Note: This approach can also be applied to individual states of large countries. For example, the approach can be applied to California or Texas or any state in the United States using the state's specific characteristics.

[763] Note: According to U.S. EIA coal contributes to less than 20% electricity production in the United States. However, it produces about 50% of the greenhouse gas emissions. It also causes the most air pollution.

[764] Note: Future potential exists for further greenhouse gas reductions in subsequent phases from natural gas power plants by including a capture and storage component.

[765] Note: Cost effective deployment of utility-scale solar and wind onshore means limiting the deployment to the level where the total lifetime cost is reasonably competitive with existing electricity costs and cost-effective deployment of hydropower and geothermal means limiting the deployment to suitable locations.

[766] 2021 Grid electricity emissions, January 2022. https://www.carbonfootprint.com/docs/2022_01_emissions_factors_sources_for_2021_electricity_v10.pdf

[767] Note: Germany has a carbon intensity of electricity generation of about 310 gCO_2 per kWh. United States has a carbon intensity of electricity generation of about 380 gCO_2 per kWh, while Australia and India have a carbon intensity that is above 650 gCO_2 per kWh. The carbon intensity of electricity generation in China is above 500 gCO_2 per kWh.

[768] Note: Climate enthusiasts, who are affluent must advance the options that are cost-inefficient and/or inconvenient in the early phases. Energy policies should not be used for this.

[769] Note: For example, targeting a warming of less than 1.5°C will require energy policies that will be unable to satisfy the critical requirements and therefore will not be sustainably supported by the global population. This will very likely lead to a poor outcome, i.e., lead to a warming above 3°C. Instead, a target that is closer to 2°C is more likely to be achieved by policies that will be sustainably supported by the global population.

[770] List of ~200 supporting scientific organizations. http://www.opr.ca.gov/facts/list-of-scientific-organizations.html

Printed in Great Britain
by Amazon